# Problems
# Fluid Flow

**Donald J. Brasch**
**Derek Whyman**
*University of Otago*
*Dunedin New Zealand*

© D. J. Brasch and D. Whyman 1986

First published in Great Britain 1986 by
Edward Arnold (Publishers) Ltd, 41 Bedford Square, London WC1B 3DQ

Edward Arnold (Australia) Pty Ltd, 80 Waverley Road,
Caulfield East, Victoria 3145, Australia

Edward Arnold, 3 East Read Street, Baltimore, Maryland 21202, U.S.A.

**British Library Cataloguing in Publication Data**

Brasch, D. J.
   Problems in fluid flow.
   I. Title  II. Whyman, D.
   532'.051    QC151

   ISBN  0-7131-3554-9

All rights reserved. No part of this publication may be reproduced,
stored in a retrieval system, or transmitted in any form or by any means,
electronic, photocopying, recording, or otherwise, without the prior
permission of Edward Arnold (Publishers) Ltd.

Text set in 10/12 Times, Monophoto
by Macmillan India Ltd, Bangalore 25.
Printed and bound in Great Britain by
J. W. Arrowsmith Ltd, Bristol.

# Preface

There are excellent texts which deal with the theory of fluid flow. However, there are fewer books which purport to illustrate application of this theory to every-day technical problems. Since, to avoid expense, little space can usually be set aside for problem solving in texts which are primarily theoretical, we seek with this book to assist in the further understanding of fluid flow systems after some degree of familiarity with basic theory has been gained from standard texts.

References in this work are to discussions of basic theory needed to solve particular problems. Usually, these references are not to original literature, but instead, they point the way to discussions in readily available textbooks.

In spite of the title of this book, perusal of the contents list will indicate that these pages were written primarily for chemical and mineral engineers. However, much of the subject matter of fluid flow is common to many courses in fluid flow, so hopefully, other engineering and applied science students may also find much subject matter of interest to them here.

The selection of symbols is always a vexed question, particularly because the English and Greek alphabets cannot provide between them sufficient symbols to represent all parameters uniquely. We have aimed to be consistent, but for the sake of clarity, we define all symbols used in each chapter in a list at the end of that chapter. Each symbol has a meaning which is unique within each chapter.

Although it is undoubtedly good experience to have to seek out data for oneself, it is not convenient to have to do this while struggling to understand a problem. So, all data required for calculations in this text are given in the statement of each problem.

Finally, in energy balance equations used in this book, we have adopted the sign convention that is used in most engineering texts. We apologise to industrial chemists who may prefer to use another sign convention.

D. J. B.<br>D. W.

# Contents

| | | |
|---|---|---|
| **Preface** | | iii |
| **1** | **Pipe flow of liquids** (D. W.) | 1 |
| 1.1 | Worked examples | 1 |
| 1.1.1 | Laminar flow, turbulent flow and the pipe-flow Reynolds number | 1 |
| 1.1.2 | Conditions in a pipeline while liquid passes in steady motion through it | 2 |
| 1.1.3 | Laminar flow and the Hagen–Poiseuille equation | 4 |
| 1.1.4 | Velocity distribution in fluid in laminar motion in a pipe | 6 |
| 1.1.5 | Comparison of laminar and turbulent flow | 7 |
| 1.1.6 | Power required for pumping, local pressure in a pipeline, and the effects on both of an increase in pipe roughness | 9 |
| 1.1.7 | Power required for pumping when the pipe system contains fittings and other resistances to flow. Assessment of the need for cleaning | 12 |
| 1.1.8 | Fluid flow rate and use of a friction group ($\phi Re^2$) versus $Re$ chart | 15 |
| 1.1.9 | Time taken to drain a tank and the use of $\phi Re^2$ versus $Re$, and $\phi$ versus $Re$ charts | 17 |
| 1.1.10 | Minimum pipe diameter to obtain a given fluid flow rate without pumping | 21 |
| 1.2 | Student Exercises | 22 |
| 1.3 | References | 25 |
| 1.4 | Notation | 26 |
| **2** | **Pipe flow of gases and gas–liquid mixtures** (D. W.) | 27 |
| 2.1 | Worked examples | 27 |
| 2.1.1 | Gas flow through a pipeline when compressibility must be considered | 27 |
| 2.1.2 | Flow of an ideal gas at maximum velocity under isothermal and adiabatic conditions | 29 |
| 2.1.3 | Flow of a non-ideal gas at maximum velocity under adiabatic conditions | 35 |

| | | |
|---|---|---|
| 2.1.4 | Venting of gas from a pressure vessel | 39 |
| 2.1.5 | Gas-flow measurement with a Venturi meter | 41 |
| 2.1.6 | Pressure drop required for flow of a gas–liquid mixture through a pipe | 44 |
| 2.2 | Student Exercises | 45 |
| 2.3 | References | 47 |
| 2.4 | Notation | 48 |

| | | |
|---|---|---|
| **3** | **Velocity boundary layers** (D. W.) | 50 |
| 3.1 | Worked examples | 50 |
| 3.1.1 | Streamline flow over a flat plate | 50 |
| 3.1.2 | Turbulent flow over a flat plate | 52 |
| 3.1.3 | Streamline and turbulent flow through a pipe, and equations of the universal velocity profile | 55 |
| 3.1.4 | Streamline flow through a pipe when flow patterns are not fully developed | 60 |
| 3.2 | Student Exercises | 62 |
| 3.3 | References | 63 |
| 3.4 | Notation | 64 |

| | | |
|---|---|---|
| **4** | **Flow measurement** (D. W.) | 65 |
| 4.1 | Worked examples | 65 |
| 4.1.1 | Use of a Pitot tube to measure flow rate in fluid streams in fully developed flow | 65 |
| 4.1.2 | Use of a Pitot tube to determine flow rate in a gas stream in which flow patterns are not fully developed | 67 |
| 4.1.3 | Use of an orifice plate and manometer to measure fluid flow rate in horizontal and vertical pipelines | 70 |
| 4.1.4 | Determination of orifice size required to measure a flow rate of fluid with an orifice meter, and comparison of pressure drops produced by orifice and venturi meters | 72 |
| 4.1.5 | Use of a rotameter to measure fluid flow rates | 74 |
| 4.1.6 | Mass of a float required to measure fluid flow rate with a rotameter | 77 |
| 4.2 | Student Exercises | 79 |
| 4.3 | References | 81 |
| 4.4 | Notation | 81 |

| | | |
|---|---|---|
| **5** | **Flow and flow measurement in open channels** (D. W.) | 83 |
| 5.1 | Worked examples | 83 |
| 5.1.1 | Use of the Manning and Chézy formulae to analyse steady, uniform flow | 83 |

Contents   vii

| | | |
|---|---|---|
| 5.1.2 | Stream depth in a channel of trapezoidal cross-section | 85 |
| 5.1.3 | Optimum base angle for a V-shaped channel. Slope of a channel | 86 |
| 5.1.4 | Stream depth for maximum velocity and for maximum volumetric flow rate in a pipe | 88 |
| 5.1.5 | Flow measurement with sharp-crested weirs | 91 |
| 5.1.6 | Equation for specific energy and analysis of streams in tranquil and shooting flow | 93 |
| 5.1.7 | Alternative depths of streams, and gradients of mild and steep slopes | 96 |
| 5.1.8 | Critical flow conditions in channels of non-rectangular section | 97 |
| 5.1.9 | Flow measurement with broad-crested weirs | 99 |
| 5.1.10 | Gradually varied flow behind a weir | 102 |
| 5.1.11 | Analysis of a hydraulic jump | 105 |
| 5.2 | Student Exercises | 109 |
| 5.3 | References | 112 |
| 5.4 | Notation | 112 |
| **6** | **Pumping of liquids** (D. W.) | **115** |
| 6.1 | Worked examples | 115 |
| 6.1.1 | Cavitation and its avoidance in suction pipes | 115 |
| 6.1.2 | Specific speed of a centrifugal pump, and similarity in centrifugal pump systems | 118 |
| 6.1.3 | System characteristic, theoretical and effective characteristics of a centrifugal pump, and flow rate at perfect match | 120 |
| 6.1.4 | Flow rate when centrifugal pumps operate singly and in parallel | 123 |
| 6.1.5 | Pumping with a reciprocating pump | 126 |
| 6.1.6 | Pumping with an air-lift pump | 128 |
| 6.2 | Student Exercises | 130 |
| 6.3 | References | 133 |
| 6.4 | Notation | 133 |
| **7** | **Flow through packed beds** (D. W.) | **136** |
| 7.1 | Worked examples | 136 |
| 7.1.1 | Determination of particle size and specific surface area for a sample of powder | 136 |
| 7.1.2 | Rate of flow of fluid through a packed bed | 138 |
| 7.1.3 | Determination of the pressure drop to drive fluid through a packed bed of Raschig rings, then of similar sized spheres, and the determination of total area of surface presented with the two types of packing | 140 |
| 7.2 | Student Exercises | 143 |
| 7.3 | References | 144 |
| 7.4 | Notation | 144 |

| | | |
|---|---|---|
| **8** | **Filtration (D. W.)** | 145 |
| 8.1 | Worked examples | 145 |
| 8.1.1 | Constant rate filtration in a plate-and-frame filter press when an incompressible filter cake forms. Evaluation of system parameters | 145 |
| 8.1.2 | Constant rate filtration, followed by filtration at constant pressure drop, then washing, to illustrate common filtration practice. Incompressible filter cakes | 148 |
| 8.1.3 | Determination of the characteristics of a filtration system (incompressible filter cake) with a small leaf filter, followed by manipulations to determine the performance of a large plate-and-frame filter, the latter being operated firstly under constant rate conditions, then with constant pressure drop | 150 |
| 8.1.4 | Constant pressure drop filtration of a suspension which gives rise to a compressible filter cake | 154 |
| 8.1.5 | Filtration on a rotary drum filter | 157 |
| 8.1.6 | Filtration in a centrifugal filter | 159 |
| 8.2 | Student Exercises | 162 |
| 8.3 | References | 164 |
| 8.4 | Notation | 164 |
| | | |
| **9** | **Forces on bodies immersed in fluids (D. J. B.)** | 166 |
| 9.1 | Worked examples | 166 |
| 9.1.1 | Drag forces and drag coefficients | 166 |
| 9.1.2 | Lift forces and lift coefficients | 168 |
| 9.1.3 | The calculation of particle diameters from terminal settling velocities | 171 |
| 9.1.4 | The prediction of the terminal settling velocity of a spherical particle | 173 |
| 9.1.5 | The effect of shape on the drag force experienced by non-spherical particles. Sphericity calculations | 174 |
| 9.1.6 | Estimation of the settling velocity of particles under conditions of hindered settling | 177 |
| 9.1.7 | The acceleration of a particle settling in a gravitational field | 180 |
| 9.2 | Student Exercises | 182 |
| 9.3 | References | 184 |
| 9.4 | Notation | 185 |
| | | |
| **10** | **Sedimentation and classification (D. J. B.)** | 186 |
| 10.1 | Worked examples | 186 |
| 10.1.1 | The determination of settling velocities from a single batch sedimentation test | 186 |

Contents ix

| | | |
|---|---|---|
| 10.1.2 | The estimation of the minimum area required for a continuous thickener in order to effect a given rate of separation of solids from a suspension | 188 |
| 10.1.3 | Settling velocities and the classification of materials which have different settling velocities | 192 |
| 10.1.4 | Density variations in a settling suspension | 198 |
| 10.1.5 | The determination of particle size distribution using a sedimentation method | 200 |
| 10.1.6 | The determination of the particle size distribution of a suspended solid by measurement of the mass rate of sedimentation. The sedimentation balance | 204 |
| 10.1.7 | The decanting of homogeneous suspensions to obtain particles of a given size range | 208 |
| 10.2 | Student Exercises | 210 |
| 10.3 | References | 214 |
| 10.4 | Notation | 214 |
| | | |
| **11** | **Fluidisation** (D. J. B.) | 216 |
| 11.1 | Worked examples | 216 |
| 11.1.1 | Particulate and aggregative fluidisation | 216 |
| 11.1.2 | The calculation of minimum flow rates, maximum flow rates and pressure drops in fluidised systems in which the flow is streamline | 218 |
| 11.1.3 | The calculation of flow rates in fluidised beds, when flow is not streamline, using the Ergun equation. Flow rates for the fluidisation of particles with a distribution of sizes | 221 |
| 11.1.4 | The estimation of vessel diameters and heights for fluidisation operations | 224 |
| 11.1.5 | Power required for pumping in fluidised beds | 227 |
| 11.1.6 | The wall effect in fluidised beds using narrow columns | 230 |
| 11.1.7 | The effect of particle size on the ratio of terminal velocity of particles in a fluid to the minimum fluidisation velocity | 232 |
| 11.2 | Student Exercises | 235 |
| 11.3 | References | 237 |
| 11.4 | Notation | 238 |
| | | |
| **12** | **Pneumatic conveying** (D. J. B.) | 240 |
| 12.1 | Worked examples | 240 |
| 12.1.1 | Flow patterns in pneumatic conveying. The transition from fluidised bed flow to moving bed flow | 240 |
| 12.1.2 | The prediction of choking velocity and choking voidage in a vertical transport line in which dilute phase flow occurs | 243 |

| | | |
|---|---|---|
| 12.1.3 | The prediction of the pressure drop in dilute phase horizontal pneumatic transport | 245 |
| 12.1.4 | The prediction of the pressure drop in dilute phase vertical pneumatic transport | 250 |
| 12.1.5 | The dense phase flow regime for the pneumatic transport of solids, upwards through vertical tubes without slugging | 254 |
| 12.2 | Student Exercises | 255 |
| 12.3 | References | 256 |
| 12.4 | Notation | 257 |
| **13** | **Centrifugal separation operations** (D. J. B.) | 259 |
| 13.1 | Worked examples | 259 |
| 13.1.1 | Equations for centrifugal force | 259 |
| 13.1.2 | Fluid pressure in a tubular bowl centrifuge. Maximum safe speed of rotation | 261 |
| 13.1.3 | Particle size determination of fine particles by batch centrifugation | 262 |
| 13.1.4 | Flow rates in continuous centrifugal sedimentation | 263 |
| 13.1.5 | The separation of two immiscible liquids by centrifugation | 265 |
| 13.1.6 | Cyclone separators | 266 |
| 13.1.7 | Efficiency of cyclone separators | 268 |
| 13.2 | Student Exercises | 269 |
| 13.3 | References | 272 |
| 13.4 | Notation | 272 |
| **Answers to exercises** | | 274 |
| **Appendix** | | 278 |

# 1 Pipe flow of liquids

## 1.1 Worked Examples

### 1.1.1 Laminar flow, turbulent flow and the pipe-flow Reynolds number

*Problem*

Establish whether fluid in the following systems will be in laminar or turbulent motion:
(i) a liquid of viscosity $= 6.30 \times 10^{-3}$ kg m$^{-1}$ s$^{-1}$, and density $= 1170$ kg m$^{-3}$, flowing through a pipe of 30.0 cm inside diameter at a rate of 150 000 barrels per day, and
(ii) oil of viscosity $= 5.29 \times 10^{-3}$ kg m$^{-1}$ s$^{-1}$, flowing through a pipeline of inside diameter $= 6.0$ cm at a rate of 0.32 kg s$^{-1}$.
1 barrel $= 0.142$ m$^3$

*Solution*

The value of the pipe-flow Reynolds number, $Re$ ($= \rho u d / \mu$), is the criterion by which flow pattern in a pipeline can be predicted (ref. 1, 2, 3 and 4). If $Re < 2100$, flow will be laminar. If $Re > 4000$, flow will be turbulent.

(i)  $$\text{Volumetric flow rate} = \frac{150\,000 \times 0.142}{24 \times 3600} \text{ m}^3 \text{ s}^{-1}$$

$$= 0.247 \text{ m}^3 \text{ s}^{-1}$$

So the average velocity of fluid in the pipe,

$$u = \frac{0.247}{\pi (15.0 \times 10^{-2})^2} = 3.49 \text{ m s}^{-1}$$

Thus, Reynolds number,

$$Re = \frac{1170 \times 3.49 \times 30.0 \times 10^{-2}}{6.30 \times 10^{-3}} = 1.94 \times 10^5$$

This is greater than 4000, so fluid in this system will be in *turbulent* motion.

## 2  Problems in fluid flow

(ii) Mass flow rate,
$$G = \frac{\pi d^2 u \rho}{4}$$

so the product,
$$\rho u = \frac{4G}{\pi d^2}$$

Thus,
$$Re = \frac{\rho u d}{\mu}$$

$$= \frac{4G}{\pi d^2} \cdot \frac{d}{\mu}$$

$$= \frac{4G}{\pi d \mu}$$

Substituting data,
$$Re = \frac{4 \times 0.32}{\pi \times 6 \times 10^{-2} \times 5.29 \times 10^{-3}}$$

$$= 1280$$

This is less than 2100, so fluid in this system will be in *laminar* motion.

### 1.1.2  Conditions in a pipeline while liquid passes in steady motion through it

*Problem*

If a pressure head of 24 cm of liquid per metre run of pipe is required to drive a liquid (density = 1120 kg m$^{-3}$) through a smooth, horizontal pipeline (inside diameter = 7.50 cm, length = 50 m) at a rate of 21.2 kg s$^{-1}$, calculate:
(i) the value of the Stanton–Pannell friction factor, $\phi$,
(ii) the shear stress exerted by the liquid on the wall of the pipeline,
(iii) the total shear force exerted by friction on the pipeline, and
(iv) the viscosity of the liquid.

Acceleration due to gravity = 9.81 m s$^{-2}$
For flow through smooth pipes, $\phi = 0.0396 \, Re^{-0.25}$

*Solution*

Consider a force balance on fluid moving between sections 1 and 2 in a length of pipe as shown in Figure 1.1.
The force on fluid in the pipe which derives from the pressure difference

$$= -\frac{\pi d^2 \Delta P}{4}$$

Pipe flow of liquids 3

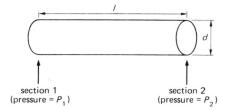

section 1 (pressure = $P_1$)    section 2 (pressure = $P_2$)

**Figure 1.1**

The force which arises from friction and opposes motion

$$= R\pi dl$$

At steady state, these forces may be equated.

Thus
$$-\frac{\pi d^2 \Delta P}{4} = R\pi dl$$

This reduces to
$$-\Delta P = 4R\frac{l}{d}$$

so that
$$\frac{-\Delta P}{\rho} = 4\left(\frac{R}{\rho u^2}\right)\frac{lu^2}{d}$$

$$= 4\phi \frac{lu^2}{d} \qquad (1.1)$$

Now, considering unit length of pipe,

$$-\Delta P = (h_2 - h_1)\rho g$$
$$= 24 \times 10^{-2} \times 1120 \times 9.81 \text{ Pa}$$
$$= 2637 \text{ Pa}$$

Also, mass flow rate, $\quad G = \dfrac{\pi d^2 u \rho}{4}$

so average fluid velocity, $\quad u = \dfrac{4G}{\pi d^2 \rho}$

$$= \frac{4 \times 21.2}{\pi (7.50 \times 10^{-2})^2 \times 1120} \text{ m s}^{-1}$$

$$= 4.28 \text{ m s}^{-1}$$

Now substituting data into equation (1.1), for unit length of pipe,

$$\frac{2637}{1120} = 4\phi \frac{1 \times 4.28^2}{7.50 \times 10^{-2}} \text{ m}^2 \text{ s}^{-2}$$

# 4 Problems in fluid flow

from which, friction factor, $\phi = \underline{0.00241}$

But, by definition, $\phi = \dfrac{R}{\rho u^2}$

So the shear stress exerted by liquid on the pipe wall,

$$R = 0.00241 \times 1120 \times 4.28^2 \, \text{N m}^{-2}$$
$$= 49.4 \, \text{N m}^{-2}$$

Now, $R$ is a force per unit area, so total shear force exerted by friction on the pipeline

$$= \pi dl R$$
$$= \pi \times 7.50 \times 10^{-2} \times 50 \times 49.4 \, \text{N}$$
$$= \underline{582 \, \text{N}}$$

Since friction factor, $\phi$, has been evaluated and the pipe may be considered smooth, Reynolds number for fluid flow in the system may be obtained either from a chart of $\phi$ versus $Re$, or by using the Blasius equation,

$$\phi = 0.0396 \, Re^{-0.25}$$

Using the Blasius equation,

$$Re^{0.25} = \dfrac{0.0396}{0.00241}$$

from which, $Re = 7.29 \times 10^4$

But, $Re = \dfrac{\rho u d}{\mu}$

So, viscosity of the liquid,

$$\mu = \dfrac{1120 \times 4.28 \times 7.50 \times 10^{-2}}{7.29 \times 10^4} \, \text{kg m}^{-1}\text{s}^{-1}$$
$$= \underline{4.93 \times 10^{-3} \, \text{kg m}^{-1}\text{s}^{-1}}$$

## 1.1.3 Laminar flow and the Hagen–Poiseuille equation

*Problem*

A flow meter consists of a horizontal, narrow tube (length = 3.50 cm, inside diameter = 1.25 mm) across which is attached a manometer. Calculate the maximum volumetric flow rate (in $\text{cm}^3 \, \text{s}^{-1}$) at which cyclohexane (density = 779 $\text{kg m}^{-3}$, viscosity = $1.02 \times 10^{-3}$ $\text{kg m}^{-1}\text{s}^{-1}$) may be passed in laminar flow through the device. Then find the difference (in cm) in interface levels in the manometer at this flow rate if the manometer contains cyclohexane over water.

Density of water = 999 kg m$^{-3}$
Acceleration due to gravity = 9.81 m s$^{-2}$

*Solution*

The maximum pipe flow Reynolds number at which stable laminar flow will obtain (ref. 1, 2, 3 and 4), = 2100,

i.e. $\qquad \rho u d/\mu = 2100$

So the greatest average velocity of cyclohexane at which laminar flow will obtain,

$$u = \frac{2100 \times 1.02 \times 10^{-3}}{779 \times 1.25 \times 10^{-3}} \text{ m s}^{-1}$$

$$= 2.20 \text{ m s}^{-1}$$

This corresponds to a volumetric flow rate

$$= \frac{\pi d^2 u}{4}$$

$$= \frac{\pi (1.25 \times 10^{-3})^2 \times 2.20}{4} \text{ m}^3 \text{ s}^{-1}$$

$$= \underline{2.70 \text{ cm}^3 \text{ s}^{-1}}$$

Now with laminar flow, the Hagen–Poiseuille equation (ref. 5, 6 and 7) will apply,

i.e. $\qquad -\Delta P = \dfrac{32 \mu u l}{d^2}$

Thus, the greatest pressure difference which can be expected to give rise to laminar flow,

$$-\Delta P = \frac{32 \times 1.02 \times 10^{-3} \times 2.20 \times 3.50 \times 10^{-2}}{(1.25 \times 10^{-3})^2} \text{ Pa}$$

$$= 1608 \text{ Pa}$$

If this gives rise to a difference in interface levels, $\Delta z$, (ref. 8, 9 and 10) in the manometer,

$$1608 = \Delta z (999 - 779) \times 9.81$$

from which, $\qquad \underline{\Delta z = 74.5 \text{ cm}}$

### 1.1.4 Velocity distribution in fluid in laminar motion in a pipe

*Problem*

Glycerol is pumped from storage tanks to rail cars through a single pipeline (inside diameter = 5.00 cm, length = 12 m) which must be used for both pure and technical grades of glycerol. After the line has been used for technical material, how much pure glycerol (in m$^3$) must be pumped through it before the liquid which appears at the end of the pipeline will contain not more than 2% of technical material? Glycerol is pumped at a rate which gives rise to laminar flow, and the two grades of glycerol may be considered to have identical densities and viscosities.

*Solution*

Glycerol appearing at the end of the pipeline will contain 2% of technical glycerol when the core of pure material occupies 98% of the cross-sectional area of the bore of the pipe at the discharge end. Let the radius of this core of pure material = $s$.

Then,
$$\pi s^2 = \frac{98}{100} \pi r^2$$

so that
$$s^2 = 0.98 r^2$$

Let the velocity of fluid in this position = $\dot{u}_x$.

Then,
$$\dot{u}_x = 2u(1 - s^2/r^2) \qquad \text{(ref. 5, 11 and 12)}$$

Now volumetric flow rate
$$= \pi r^2 u$$

$$= \frac{\pi r^2 \dot{u}_x}{2(1 - s^2/r^2)}$$

If the operation takes $t$ seconds,
$$\dot{u}_x = 12/t \text{ m s}^{-1}$$

Then the volume of pure glycerol which must be pumped in $t$ seconds to cause 2% of technical material only to appear at the discharge end

$$= \frac{\pi (2.50 \times 10^{-2})^2 \times 12}{2[1 - (0.98 r^2)/r^2]} \text{ m}^3$$

$$= \underline{0.589 \text{ m}^3}$$

## 1.1.5 Comparison of laminar and turbulent flow

### Problem

At what percentage of a pipe's diameter must a Pitot tube be located in from the walls of a pipeline to indicate directly the average fluid velocity
(i) when fluid is in streamline motion, and
(ii) when fluid is in turbulent flow?
(iii) What would be the percentage error incurred if a Pitot tube were located at the position at which it would indicate average fluid velocity directly for turbulent flow if streamline flow were actually to obtain?

### Solution

(i) For streamline flow, velocity of fluid at a distance $s$ from the centre of a pipe, (see ref. 12, 13 and 14)

$$\dot{u}_x = -\frac{1}{4\mu} \cdot \frac{dP}{dl}(r^2 - s^2) \qquad (1.2)$$

So, axial fluid velocity,

$$\dot{u}_s = -\frac{1}{4\mu} \cdot \frac{dP}{dl} r^2$$

But for streamline flow, average fluid velocity,

$$u = \frac{1}{2}\dot{u}_s \qquad \text{(ref. 5, 7 and 11)}$$

$$= -\frac{1}{2} \cdot \frac{1}{4\mu} \cdot \frac{dP}{dl} r^2 \qquad (1.3)$$

So where $\quad u = \dot{u}_x$

from equations (1.2) and (1.3),

$$r^2 - s^2 = \frac{1}{2}r^2$$

from which, $\quad s = \dfrac{r}{2^{0.5}}$

But, $\quad s = r - y$

So, $\quad y = r\left(1 - \dfrac{1}{2^{0.5}}\right)$

$$= \frac{d}{2}\left(1 - \frac{1}{2^{0.5}}\right)$$

and $\quad \dfrac{y}{d} = 0.1464$

So for streamline flow, a Pitot tube would have to be located at 14.64% of the pipe's diameter in from the pipe wall in order to indicate average fluid velocity directly.

(ii) For turbulent flow, fluid velocity at a distance $y$ from the pipe wall,

$$\dot{u}_x = \dot{u}_s \left(\frac{y}{r}\right)^{\frac{1}{7}} \qquad \text{(ref. 15 and 16)}$$

Also, average fluid velocity,

$$u = \frac{49}{60} \dot{u}_s \qquad \text{(ref. 17 and 18)}$$

So when $\dot{u}_x = u$,

$$\left(\frac{y}{r}\right)^{\frac{1}{7}} = \frac{49}{60}$$

so that

$$\frac{y}{r} = \left(\frac{49}{60}\right)^7$$

and

$$\frac{y}{d} = \frac{1}{2}\left(\frac{49}{60}\right)^7 = 0.1211$$

So for turbulent flow, a Pitot tube would have to be located at 12.1% of the pipe's diameter in from the pipe wall in order to indicate average fluid velocity directly.

(iii) Since for turbulent flow,

$$\frac{y}{d} = 0.1211$$

$$y = 0.1211 \times 2 \times r = 0.2422\,r$$

But $\quad s = r - y$

So $\quad s = r(1 - 0.2422) = 0.7578\,r$

Inserting this value for $s$ into equation (1.2)

$$\dot{u}_x = -\frac{1}{4\mu} \cdot \frac{dP}{dl} \cdot r^2(1 - 0.7578^2)$$

But actually, average velocity,

$$u = -\frac{1}{2} \cdot \frac{1}{4\mu} \cdot \frac{dP}{dl} r^2 \qquad \text{(see equation (1.3))}$$

So the fractional error incurred by mis-locating the Pitot tube

$$= \frac{\dot{u}_x - u}{u}$$

$$= \frac{(1 - 0.7578^2) - \frac{1}{2}}{\frac{1}{2}}$$

$$= -0.1485$$

Thus, the Pitot tube would indicate an average value for the velocity which was 14.9% low if streamline flow were actually to obtain.

### 1.1.6 Power required for pumping, local pressure in a pipeline, and the effects on both of an increase in pipe roughness

*Problem*

Aqueous nitric acid (containing 12% by weight of $HNO_3$) is to be pumped from a storage tank, through 278 m of steel pipe (inside diameter = 3.20 cm, absolute roughness when new = 0.0035 mm) to discharge freely at a point 57.4 m above the pump outlet. The pump will be situated immediately beside the storage tank and will draw acid from a point 5.60 m below the surface of liquid in the tank. Calculate:
(i) the power required to deliver acid solution at a rate of 2.35 kg s$^{-1}$,
(ii) the gauge pressure which will be indicated on a gauge at the pump outlet,
(iii) the percentage increase in power which the pump must be designed to deliver to maintain flow at 2.35 kg s$^{-1}$ when corrosion has increased the absolute roughness of the pipeline to 0.050 mm, and
(iv) the gauge pressure at the pump outlet when pipe roughness has increased to 0.050 mm.

Density of the nitric acid = 1068 kg m$^{-3}$
Viscosity of the nitric acid = $1.06 \times 10^{-3}$ kg m$^{-1}$ s$^{-1}$
Acceleration due to gravity = 9.81 m s$^{-2}$

*Solution*

(i) The system is shown in Figure 1.2. Firstly, applying the overall mechanical energy balance equation (ref. 19, 20, 21 and 22):

$$\Delta\left(\frac{u^2}{2\alpha}\right) + g\Delta z + \frac{\Delta P}{\rho} + W_s + 4\phi\frac{lu^2}{d} = 0 \qquad (1.4)$$

to the system between the surface of acid in the storage tank (section 1) and the point of discharge (section 2), since the pressure at sections 1 and 2 is atmospheric, $\Delta P = 0$. So the shaft work required for pumping then becomes:

$$-W_s = \Delta\left(\frac{u^2}{2\alpha}\right) + g\Delta z + 4\phi\frac{lu^2}{d} \qquad (1.5)$$

# 10 Problems in fluid flow

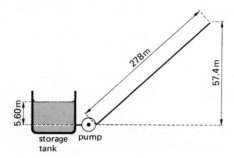

**Figure 1.2**

To find the average fluid velocity, $u$, the volumetric flow rate of acid

$$= \frac{2.35}{1068} \text{ m}^3 \text{ s}^{-1}$$

Since the cross-sectional area of the pipe

$$= \frac{\pi (3.20 \times 10^{-2})^2}{4} \text{ m}^2$$

$$u = \frac{2.35 \times 4}{1068 \times \pi (3.20 \times 10^{-2})^2} \text{ m s}^{-1}$$

$$= 2.74 \text{ m s}^{-1}$$

Now,
$$Re = \frac{1068 \times 2.74 \times 3.20 \times 10^{-2}}{1.06 \times 10^{-3}}$$

$$= 8.83 \times 10^4$$

So flow will be turbulent, and $\alpha \approx 1$ (ref. 23, 24 and 25).

Now, the relative roughness, $e/d$, must be evaluated in order to read off a value for $\phi$ from a chart of $\phi$ versus $Re$. When the pipe is new,

$$\frac{e}{d} = \frac{0.0035 \times 10^{-3}}{3.20 \times 10^{-2}}$$

$$= 1.09 \times 10^{-4}$$

So reading from a chart of $\phi$ versus $Re$,

$$\phi = 0.00225$$

Now
$$\Delta z = 57.4 - 5.60 = 51.8 \text{ m}$$

Substituting data into equation (1.5), when the pipe is new,

$$-W_s = \left(\frac{2.74^2}{2} - \frac{0^2}{2}\right) + 9.81 \times 51.8 + \frac{4 \times 0.00225 \times 278 \times 2.74^2}{3.20 \times 10^{-2}} \text{ J kg}^{-1}$$

$$= 1099 \text{ J kg}^{-1}.$$

Since the mass flow rate is to be 2.35 kg s$^{-1}$, the power required to pump acid when the pipeline is new

$$= 1099 \times 2.35 \text{ J s}^{-1}$$
$$= \underline{2.58 \text{ kW}}$$

(ii) Gauge pressure is pressure above atmospheric. So to find the gauge pressure, $P'_2$, indicated at the pump outlet, equation (1.4) must be applied between the surface of the acid (section 1) in the storage tank, and the pump outlet (section 2).

Then
$$\Delta P = (P'_2 + P_{atm}) - P_{atm}$$
$$= P'_2$$
$$\Delta z = 0 - 5.60 \text{ m} = -5.60 \text{ m}$$

The length of pipe, $l$, through which acid flows to section 2 may be considered negligibly small. So considering equation (1.4),

$$\frac{P'_2}{\rho} = -\Delta\left(\frac{u^2}{2\alpha}\right) - g\Delta z - W_s - 4\phi\frac{lu^2}{d} \qquad (1.6)$$

which on substituting data becomes:

$$P'_2 = -\frac{2.74^2 \times 1068}{2} + (9.81 \times 5.60 \times 1068) + (1099 \times 1068) \text{ Pa}$$

$$= 1228 \text{ kPa}$$

Thus, gauge pressure at the pump outlet when piping is new = $\underline{1228 \text{ kPa}}$.

(iii) When pipe roughness has increased to 0.050 mm,

$$\frac{e}{d} = \frac{0.050 \times 10^{-3}}{3.20 \times 10^{-2}} = 1.56 \times 10^{-3}$$

So referring back to part (i) and to a chart of $\phi$ versus $Re$, when

$$Re = 8.83 \times 10^4$$
$$\phi = 0.0029$$

Applying equation (1.4) between the surface of acid in the storage tank (section 1) and the pipe outfall (section 2) again,

$$-W_s = \frac{2.74^2}{2} + 9.81 \times 51.8 + \frac{4 \times 0.0029 \times 278 \times 2.74^2}{3.20 \times 10^{-2}} \text{ J kg}^{-1}$$

$$= 1268 \text{ J kg}^{-1}$$

Since the mass flow rate is still to be 2.35 kg s$^{-1}$, the power required to pump fluid when the pipeline has become corroded

$$= 1268 \times 2.35 \text{ J s}^{-1} = 2.98 \text{ kW}$$

The power increase which must be allowed for, therefore,

$$= \frac{2.98 - 2.58}{2.58} \times 100\%$$

$$= 15.5\%$$

(iv) The gauge pressure, $P'_2$, which will be observed at the pump outlet after corrosion may now be deduced by substituting appropriate data into equation (1.6). Thus:

$$P'_2 = -\frac{2.74^2 \times 1068}{2} + (9.81 \times 5.60 \times 1068) + (1268 \times 1068) \text{ Pa}$$

$$= 1409 \text{ kPa}$$

i.e. gauge pressure which will be observed at the pump outlet after corrosion = 1409 kPa.

### 1.1.7 The power required for pumping when the pipe system contains fittings and other resistances to flow. Assessment of the need for cleaning

*Problem*

Water, at 46 °C, is to be pumped from a thermal spring at a rate of 4800 dm$^3$ h$^{-1}$, in a pipe (inside diameter when clean = 3.8 cm), through a distance of 155 m in a horizontal direction, then vertically upwards through a height of 10.5 m. It will discharge freely from the end of the pipeline, and pressure at each end of the pipeline is approximately atmospheric.

In the pipe, there is a control valve which, in its normal position, has a resistance to flow, equivalent to 200 pipe diameters, and other fittings equivalent to 90 pipe diameters. There is also a heat exchanger near the end of the line and across which there is a decrease in pressure head of 154 cm of water.

If the pipe has an absolute roughness of $1.8 \times 10^{-2}$ cm when clean, what power will be required to drive the pump if the pump is 60% efficient? If the pump is situated beside the spring, and the gauge pressure at the pump outlet is 219 kPa, establish whether the pipeline should be cleaned.

Density of water at 46 °C = 990 kg m$^{-3}$
Viscosity of water at 46 °C = $5.88 \times 10^{-4}$ kg m$^{-1}$ s$^{-1}$
Acceleration due to gravity = 9.81 m s$^{-2}$

## Solution

Here, the overall mechanical energy balance equation (ref. 19, 20, 21 and 22) may be applied to the system between the surface of water in the spring (section 1) and the point of discharge from the pipeline (section 2). The equation may be written:

$$\Delta\left(\frac{u^2}{2\alpha}\right) + \frac{\Delta P}{\rho} + g\Delta z + W_s + 4\phi\frac{lu^2}{d} = 0 \qquad (1.7)$$

Because pressure at each end of the pipeline is approximately atmospheric,

$$\Delta P = 0.$$

Furthermore, $\Delta z = 10.5 - 0 = 10.5$ m

The flow rate required $= 4800 \text{ dm}^3 \text{ h}^{-1}$

$$= \frac{4800 \times 10^{-3}}{3600} = 1.333 \times 10^{-3} \text{ m}^3 \text{ s}^{-1}$$

Because the pipe, when clean, has an inside diameter = 3.8 cm, the average velocity of water in the pipeline then

$$= \frac{1.333 \times 10^{-3} \times 4}{\pi (3.8 \times 10^{-2})^2} \text{ m s}^{-1}$$

$$= 1.176 \text{ m s}^{-1}$$

So the pipe-flow Reynolds number for flow in the system,

$$Re = \frac{990 \times 1.176 \times 3.8 \times 10^{-2}}{5.88 \times 10^{-4}}$$

$$= 7.52 \times 10^4$$

So flow will be turbulent.
When the pipe is clean, its relative roughness,

$$\frac{e}{d} = \frac{1.8 \times 10^{-2}}{3.8} = 0.004\,74$$

So reading from a chart of $\phi$ versus $Re$

$$\phi = 0.0038$$

Furthermore, because flow is turbulent, $\alpha \approx 1$, (ref. 23, 24 and 25)

and $$\Delta\left(\frac{u^2}{2\alpha}\right) = \frac{1.176^2}{2} = 0.691 \text{ m}^2 \text{ s}^{-2}$$

Now considering the piping system, the effective length of the pipeline will be greater than actual because of the effects of pipe fittings. Thus, the effective

## 14 Problems in fluid flow

length of the pipe when clean

$$= 155 + 10.5 + 200(3.8 \times 10^{-2}) + 90(3.8 \times 10^{-2}) \text{ m}$$
$$= 176.5 \text{ m}$$

Finally, with a decrease of pressure head across the heat exchanger, energy is dissipated there too. Energy lost per unit mass of water flowing through the heat exchanger

$$= g\Delta h$$
$$= 9.81 \times 154 \times 10^{-2} = 15.11 \text{ m}^2 \text{ s}^{-2}$$

So substituting data into equation (1.7), and including energy lost through the heat exchanger, the work required from the pump for each kilogram of water,

$$-W_s = 0.691 + (9.81 \times 10.5) + \frac{4 \times 0.0038 \times 176.5 \times 1.176^2}{3.8 \times 10^{-2}} + 15.11 \text{ m}^2 \text{ s}^{-2}$$

$$= 216.4 \text{ J kg}^{-1}$$

Now the mass flow rate of water required

$$= \frac{4800 \times 10^{-3} \times 990}{3600} = 1.32 \text{ kg s}^{-1}$$

So the power it is necessary to transfer to the water stream

$$= 216.4 \times 1.32 \text{ J s}^{-1} = 0.286 \text{ kW}$$

Since the pump is only 60% efficient, the power required to drive the pump when the pipeline is clean

$$= 0.286 \times \frac{100}{60} \text{ kW}$$

$$= \underline{0.477 \text{ kW}}$$

To find the gauge pressure to be expected at the pump outlet when the pipe is clean, the overall energy balance equation must now be applied between the surface of water in the spring (section 1) and the pump outlet (section 2). Now if $P'_2$ represents gauge pressure at section 2,

$$\Delta P = (P'_2 + P_{atm}) - P_{atm}$$
$$= P'_2$$

Also, in passing from section 1 to section 2, both the elevation imposed on the liquid, and the length of pipe traversed, may be considered negligibly small. So substituting relevant data into equation (1.7),

$$\frac{P'_2}{990} = -0.691 + 216.4$$

from which
$$P'_2 = 213.6 \text{ kPa}$$

Since the gauge actually indicates a pressure of 219 kPa above atmospheric, observed pressure is only slightly above the pressure to be expected when the pipeline is clean. So, the pipeline does not need to be cleaned.

### 1.1.8 Fluid flow rate and use of a friction group ($\phi Re^2$) versus $Re$ chart

*Problem*

Oil in a storage tank is loaded into a tanker through a horizontal pipeline (inside diameter = 10.5 cm, length = 87.0 m). If the pipe entrance is initially 16.8 m below the surface of oil in the storage tank, calculate the initial volumetric flow rate at which oil will flow into the tanker,
(i) when the oil is draining freely through the pipeline, and
(ii) when flow is accelerated by a pump injecting 320 J of energy per kilogram of oil into the fluid stream.
Changes in kinetic energy in the oil stream may be neglected.

Density of the oil = 908 kg m$^{-3}$
Viscosity of the oil = 3.90 × 10$^{-2}$ kg m$^{-1}$ s$^{-1}$
Acceleration due to gravity = 9.81 m s$^{-2}$
Absolute roughness of the pipe = 0.046 mm

*Solution*

(i) To find volumetric flow rate, the average fluid velocity, $u$, is required. However, this cannot be found directly by using the overall mechanical energy balance equation because $u$ is required to calculate $Re$, from which to evaluate friction factor, $\phi$.

Since changes in kinetic energy may be neglected, we may consider resistance to flow to occur solely in the pipeline. Then when oil is draining freely, pressure drop due to the fluid head will be the only factor causing flow in the pipeline. Now pressure drop, $-\Delta P$, can be calculated. So friction group, $\phi Re^2$, may be evaluated because:

$$\phi Re^2 = \frac{R}{\rho u^2} \cdot \left(\frac{\rho u d}{\mu}\right)^2$$

$$= \frac{R \rho d^2}{\mu^2}$$

Then because pressure forces only have to overcome friction in the pipeline,

$$-\frac{\Delta P \pi d^2}{4} = \pi d l R \qquad \text{(see section 1.1.2)}$$

# 16 Problems in fluid flow

so that
$$R = -\frac{\Delta P \cdot d}{4l}$$

and
$$\phi Re^2 = -\frac{\Delta P \rho d^3}{4l\mu^2} \tag{1.8}$$

Consider the whole pipeline,
$$\Delta P = P_{atm} - [(16.8 \times 908 \times 9.81) + P_{atm}] \text{ Pa}$$
$$= -149.6 \text{ kPa}$$

So
$$\phi Re^2 = \frac{149.6 \times 10^3 \times 908 \times (10.5 \times 10^{-2})^3}{4 \times 87.0 \times (3.90 \times 10^{-2})^2}$$
$$= 2.97 \times 10^5$$

Now if relative roughness, $e/d$, is evaluated, the pipe-flow Reynolds number for the system may be obtained from a chart of $\phi Re^2$ versus $Re$ (see Appendix).

Thus,
$$\frac{e}{d} = \frac{0.046 \times 10^{-3}}{10.5 \times 10^{-2}} = 4.38 \times 10^{-4}$$

so that
$$Re = 8.0 \times 10^3$$

Now
$$Re = \frac{\rho u d}{\mu}$$

so the average velocity of the oil in the pipeline
$$u = \frac{8.0 \times 10^3 \times 3.90 \times 10^{-2}}{908 \times 10.5 \times 10^{-2}} \text{ m s}^{-1}$$
$$= 3.27 \text{ m s}^{-1}$$

and the initial volumetric flow rate at which the oil will drain into the tanker
$$= \frac{\pi (10.5 \times 10^{-2})^2 \times 3.27}{4} \text{ m}^3 \text{ s}^{-1}$$
$$= \underline{2.83 \times 10^{-2} \text{ m}^3 \text{ s}^{-1}}$$

(ii) The increase in pressure drop when the pump is working
$$= -W_s \times \rho$$
$$= 320 \times 908 \text{ Pa} = 290.6 \text{ kPa}$$

In this case, therefore, the effective pressure difference required in equation (1.8)
$$= -149.6 - 290.6 = -440.2 \text{ kPa}$$

So $$\phi Re^2 = \frac{440.2 \times 10^3 \times 908 \times (10.5 \times 10^{-2})^3}{4 \times 87.0 \times (3.90 \times 10^{-2})^2}$$

$$= 8.74 \times 10^5$$

Now since $\dfrac{e}{d} = 4.38 \times 10^{-4}$

from a chart of $\phi Re^2$ versus $Re$,

$$Re = 1.50 \times 10^4$$

So as before, $$u = \frac{1.50 \times 10^4 \times 3.90 \times 10^{-2}}{908 \times 10.5 \times 10^{-2}} \text{ m s}^{-1}$$

$$= 6.14 \text{ m s}^{-1}$$

and initial volumetric flow rate $$= \frac{\pi (10.5 \times 10^{-2})^2 \times 6.14}{4} \text{ m}^3 \text{ s}^{-1}$$

$$= 5.32 \times 10^{-2} \text{ m}^3 \text{ s}^{-1}$$

## 1.1.9 Time taken to drain a tank and the use of $\phi Re^2$ versus $Re$, and $\phi$ versus $Re$ charts

*Problem*

Water is allowed to drain from a cylindrical tank (inside diameter = 28.0 cm) through a vertical tube (inside diameter = 0.42 cm, length = 52.0 cm) fixed into the circular base of the tank. If the tube has an absolute roughness of $1.2 \times 10^{-3}$ cm, and the friction factor for fluid flow in the system is considered to remain approximately constant, calculate how long it will take for the water level in the tank to fall 20 cm if water in the tank is initially 45 cm deep, and fully developed flow obtains throughout the drain tube. Explain why the time for drainage which would be observed experimentally would be greater than the time calculated.

Density of the water = 1000 kg m$^{-3}$
Viscosity of the water = $1.25 \times 10^{-3}$ kg m$^{-1}$ s$^{-1}$
Acceleration due to gravity = 9.81 m s$^{-2}$

*Solution*

The overall mechanical energy balance equation:

$$\Delta \left(\frac{u^2}{2\alpha}\right) + \frac{\Delta P}{\rho} + g\Delta z + W_s + 4\phi \frac{lu^2}{d} = 0$$

(ref. 19, 20, 21 and 22)

18  *Problems in fluid flow*

may be applied to an energy balance between sections 1 and 2 in the system shown in Figure 1.3.

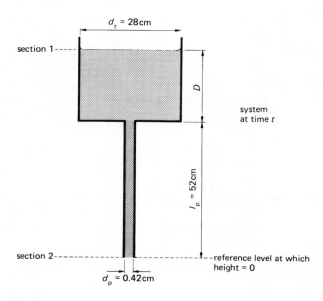

**Figure 1.3**

If the water level falls by d$D$ in time d$t$,

$$u_1 = -\frac{dD}{dt}$$

Thus, volumetric flow rate $= -\pi(14.0 \times 10^{-2})^2 \cdot \dfrac{dD}{dt}$ m$^3$ s$^{-1}$

So
$$u_2 = -\frac{\pi(14.0 \times 10^{-2})^2}{\pi(0.21 \times 10^{-2})^2} \cdot \frac{dD}{dt} \text{ m s}^{-1}$$

$$= -4444 \times \frac{dD}{dt} \text{ m s}^{-1}$$

Thus
$$\Delta u^2 = (4444^2 - 1^2)\left(-\frac{dD}{dt}\right)^2 \text{ m}^2 \text{ s}^{-2}$$

$$= 1.975 \times 10^7 \left(-\frac{dD}{dt}\right)^2 \text{ m}^2 \text{ s}^{-2}$$

In this system, both $\Delta P$ and $W_s = 0$,
but
$$\Delta z = -(0.52 + D) \text{ m}$$

Now, friction arises from flow in two sections,

so $$F = 4\phi_t \frac{Du_1^2}{d_t} + 4\phi_p \frac{l_p u_2^2}{d_p}$$

$$= \frac{4\phi_t D}{0.28}\left(-\frac{dD}{dt}\right)^2 + \frac{4\phi_p \times 0.52 \times 4444^2}{0.42 \times 10^{-2}}\left(-\frac{dD}{dt}\right)^2 \text{ m}^2\text{s}^{-2}$$

The first term will be much smaller than the second, so may be neglected,

so that $$F = 9.78 \times 10^9 \, \phi_p \left(-\frac{dD}{dt}\right)^2 \text{ m}^2\text{s}^{-2}$$

If it is assumed that fluid flow will be turbulent, $\alpha \approx 1$ (ref. 23, 24 and 25). (The validity of this assumption will be checked later.) Then substitution of data into the energy balance equation gives:

$$\frac{1.975 \times 10^7}{2}\left(-\frac{dD}{dt}\right)^2 - 9.81(0.52 + D) + 9.78 \times 10^9 \, \phi_p \left(-\frac{dD}{dt}\right)^2 = 0$$

Rearranging, $$\left(-\frac{dD}{dt}\right)^2 = \frac{9.81(0.52 + D)}{9.875 \times 10^6 + 9.78 \times 10^9 \, \phi_p}$$

from which,
$$dt = -\left(\frac{9.875 \times 10^6 + 9.78 \times 10^9 \, \phi_p}{9.81}\right)^{0.5} (0.52 + D)^{-0.5} \, dD$$

To find the time required for the water level to fall, this equation must be integrated between $D = 0.45$ m at $t = 0$, and $D = 0.25$ m at $t = t$. Thus:

$$t = -(1.007 \times 10^6 + 9.969 \times 10^8 \, \phi_p)^{0.5} \int_{0.45}^{0.25} (0.52 + D)^{-0.5} \, dD \text{ s}$$

The integral may be evaluated by writing firstly:

$$(0.52 + D) = x$$

so that, $$dD = dx$$

Then making also appropriate changes to the limits of integration,

$$t = -(1.007 \times 10^6 + 9.969 \times 10^8 \, \phi_p)^{0.5} \int_{0.97}^{0.77} x^{-0.5} \, dx \text{ s}$$

$$= 0.2148 (1.007 \times 10^6 + 9.969 \times 10^8 \, \phi_p)^{0.5} \text{ s}$$

$\phi_p$ must now be evaluated. Since $u_2$ is not known, $\phi_p$ cannot be evaluated by first evaluating a Reynolds number.

So $$\phi_p \cdot Re_p^2 = \frac{R_p}{\rho u_2^2}\left(\frac{\rho u_2 d_p}{\mu}\right)^2$$

20  Problems in fluid flow

must be evaluated instead. Considering a force balance on the drain pipe,

$$R_p = -\Delta P_p \cdot \frac{d_p}{4l_p} \qquad \text{(cf. example 1.1.2)}.$$

Since the drain pipe is vertical, both the head of fluid, $D$, together with the column of fluid, $l_p$, actually in the drain pipe, will contribute to the effective pressure difference, $\Delta P_p$, driving fluid through the drain pipe.

Thus
$$\Delta P_p = P_{atm} - [P_{atm} + D\rho g - (-l_p \rho g)]$$
$$= -(D + l_p)\rho g$$

So
$$\phi_p \cdot Re_p^2 = (D + l_p) \frac{g\rho^2 d_p^3}{4l_p \mu^2}$$

$$= \frac{(D + 0.52) \times 9.81 \times 1000^2 \times (0.42 \times 10^{-2})^3}{4 \times 0.52 (1.25 \times 10^{-3})^2}$$

$$= 2.236 \times 10^5 (D + 0.52)$$

Now, $D$ varies from 0.45 m to 0.25 m, so $\phi_p \cdot Re_p^2$ varies from $2.169 \times 10^5$ to $1.722 \times 10^5$. The relative roughness of the pipe,

$$\frac{e}{d} = \frac{1.2 \times 10^{-3}}{0.42} = 0.00286$$

So now, from a $(\phi \cdot Re^2)$ versus $Re$ chart, $Re_p$ varies from $6.5 \times 10^3$ to $5.5 \times 10^3$. Flow in the pipe will be turbulent, therefore, and the assumption made earlier that $\alpha \approx 1$ is in fact justified.

Now from a $\phi$ versus $Re$ chart, $\phi_p$ varies from 0.0047 to 0.0048. It is approximately constant, therefore, so it is justifiable to take the average value,

$$\phi_p = 0.00475$$

Finally, therefore, the time required for the water level to fall in the tank,

$$t = 0.2148(1.007 \times 10^6 + 9.969 \times 10^8 \times 0.00475)^{0.5} \text{ s}$$
$$= 515 \text{ s}$$
$$= \underline{8 \text{ mins } 35 \text{ s}}$$

This drainage time would only obtain if there were fully developed flow conditions throughout the system. In practice, an appreciable part of the drain tube would constitute the entry length in which fluid flow would not be fully developed. Peripheral velocity gradients in this region would be greater than obtain with fully developed flow, so that retarding shear stresses actually experienced would be greater than those allowed for by assuming fully developed flow throughout the drain tube. The drainage time observed experimentally would, therefore, be greater than that calculated above.

## 1.1.10 Minimum pipe diameter to obtain a given fluid flow rate without pumping

*Problem*

Water is to be drawn from a reservoir for irrigation. The pipeline (length = 485 m, absolute roughness = $8.2 \times 10^{-3}$ cm) will discharge freely at a point 4.9 m below the surface of the reservoir. If a flow rate of 1500 gal h$^{-1}$ is required, calculate the minimum diameter of pipe which must be used if the water is to be drawn without pumping. Changes in kinetic energy of the fluid may be neglected.

Density of water = 1000 kg m$^{-3}$
Viscosity of water = $1.42 \times 10^{-3}$ kg m$^{-1}$ s$^{-1}$
1 gallon = $4.545 \times 10^{-3}$ m$^3$
Acceleration due to gravity = 9.81 m s$^{-2}$

*Solution*

Consider the overall mechanical energy balance equation (ref. 19, 20, 21 and 22):

$$\Delta\left(\frac{u^2}{2\alpha}\right) + \frac{\Delta P}{\rho} + g\Delta z + W_s + 4\phi\frac{lu^2}{d} = 0$$

Since changes in kinetic energy may be neglected, this equation will be applied between the inlet and outlet of the pipeline. Now, in this system, both $\Delta u^2$ and $W_s = 0$. Also, the volumetric flow rate,

$$Q = \frac{\pi d^2 u}{4}$$

so that
$$u = \frac{4Q}{\pi d^2} \tag{1.9}$$

Also, $\quad \Delta P = \Delta h \rho g$

Now substituting into the energy balance equation,

$$\frac{\Delta h \rho g}{\rho} + g\Delta z + \frac{4\phi l}{d}\left(\frac{4Q}{\pi d^2}\right)^2 = 0$$

from which, $\quad -d^5 g(\Delta h + \Delta z) = \dfrac{64\phi l Q^2}{\pi^2} \tag{1.10}$

Now, $\Delta h + \Delta z = -4.9$ m

and $Q = 1500$ gal h$^{-1}$

$$= \frac{1500 \times 4.545 \times 10^{-3}}{3600} \text{ m}^3\text{ s}^{-1}$$

$$= 1.894 \times 10^{-3} \text{ m}^3\text{ s}^{-1}$$

Now in order to calculate pipe diameter from equation (1.10), a value is required for $\phi$. This cannot be found in the usual way from a chart of $\phi$ versus $Re$, because $u$ is not known, and $u$ is required to calculate $Re$. Also, $Re$ cannot be found first from a chart of $\phi \cdot Re^2$ versus $Re$ because the unknown pipe diameter is required to calculate $\phi \cdot Re^2$. So, a reasonable value for $d$ must be selected, and checked for validity afterwards.

If it is assumed that pipe diameter, $d$, is $\approx 6$ cm, then using equation (1.9)

$$u = \frac{4 \times 1.894 \times 10^{-3}}{\pi \times 0.06^2} = 0.6699 \text{ m s}^{-1}$$

so that $$Re = \frac{1000 \times 0.6699 \times 0.06}{1.42 \times 10^{-3}} = 2.83 \times 10^4$$

Also, the relative roughness of the pipe,

$$\frac{e}{d} = \frac{8.2 \times 10^{-3}}{6} = 1.367 \times 10^{-3}$$

So from a chart of $\phi$ versus $Re$,

$$\phi = 0.0033$$

Now, substituting data into equation (1.10)

$$d^5 \times 9.81 \times 4.9 = \frac{64 \times 0.0033 \times 485 \, (1.894 \times 10^{-3})^2}{\pi^2}$$

from which $$d = \underline{5.995 \text{ cm}}$$

This is very close to the value of $d$ assumed earlier in the selection of a reasonable value for $\phi$, so refinement in the value of $d$ selected is not required, and the minimum diameter of pipe required = $\underline{6.00 \text{ cm}}$.

## 1.2 Student Exercises

1 Establish whether liquid sulphur, flowing in a pipe (inside diameter = 7.50 cm), will be in laminar or turbulent motion:
(i) at low temperature when its viscosity = $7.22 \times 10^{-3}$ kg m$^{-1}$ s$^{-1}$, density = 2014 kg m$^{-3}$, and when it is flowing at 2.85 dm$^3$ s$^{-1}$, and

(ii) at a higher temperature when its viscosity = 0.32 kg m$^{-1}$ s$^{-1}$, density is not known, and the flow rate has reduced to 3.85 kg s$^{-1}$.
(iii) What will be the mass flow rate of sulphur in the system of section (i) above?

2   A decrease of 71.0 cm in head of liquid is required to drive a liquid (density = 984 kg m$^{-3}$) through a smooth, horizontal pipeline (inside diameter = 55.0 cm, length = 658 m) at a volumetric flow rate of 206 dm$^3$ s$^{-1}$. Calculate:
(i) the pressure difference (in Pa) which obtains for each metre run of the pipe,
(ii) the value which obtains for the Stanton–Pannell friction factor,
(iii) the total shear force experienced by the pipeline arising from friction,
(iv) the viscosity of the liquid, and
(v) the velocity gradient which obtains in the liquid at the walls of the pipeline.
Acceleration due to gravity = 9.81 m s$^{-2}$
For flow through smooth pipes, $\phi = 0.0396\, Re^{-0.25}$

3   It is found that it takes 21 min 51 s for 200 cm$^3$ of an aqueous fluid to flow through a horizontal capillary tube (inside diameter = 1.4 mm, length = 4.8 cm) under a constant pressure drop which gives rise to a 6.00 cm difference in oil levels in a manometer coupled across the capillary tube. The leads to the manometer are filled with the aqueous fluid. If the aqueous fluid and the oil have densities of 1074 kg m$^{-3}$ and 889 kg m$^{-3}$ respectively, and laminar flow conditions may be expected in the capillary,
(i) calculate the viscosity of the aqueous fluid, then
(ii) check that laminar flow conditions actually obtained by calculating the pipe-flow Reynolds number for the system.
Acceleration due to gravity = 9.81 m s$^{-2}$

4   Pure glycerol is passed through a pipeline previously used to deliver glycerol of technical grade until only 2% of the technical material remains in the discharge end of the pipe. Calculate the volume percentage of technical material in the glycerol actually pouring from the pipe at that instant, assuming that laminar flow conditions exist in the pipeline, and that the densities and viscosities of the two grades of glycerol are both identical.

5   (i) Show that the ratio of distance from a pipe wall to pipe diameter for positions at which local fluid velocity is equal to average fluid velocity when laminar flow obtains in a pipe = 0.1464:1.
(ii) To what percentage would a measured value of average velocity of a fluid stream be in error if a Pitot tube were located at the position to indicate average velocity directly for laminar flow when flow was actually turbulent.

24 Problems in fluid flow

6   Water is drawn from a reservoir at a point 33.4 m below the water surface, passes firstly to pumps, then on through a pipeline (inside diameter = 85.0 cm, total length = 4.28 km). The pipeline rises to a point 164 m above, and 2.64 km away from the pumps. The pipeline then falls 47.2 m over the remaining distance, and there is a throttling valve at the outfall. Calculate:
(i) the power which must be supplied to the pumps if they are 68 % efficient and have to pump water at 1600 kg s$^{-1}$ to the crest of the hill through the pipeline which is initially empty,
(ii) the minimum value to which pressure may be allowed to fall at the outfall throttling valve if cavitation is to be avoided in the pipeline, and
(iii) the power which must be supplied to the pumps when the pipeline is full of water flowing at 1600 kg s$^{-1}$.

Density of water = 999 kg m$^{-3}$

Viscosity of water = $1.34 \times 10^{-3}$ kg m$^{-1}$ s$^{-1}$

Vapour pressure of water = 1.40 kPa

Atmospheric pressure on the day = 103.2 kPa

Acceleration due to gravity = 9.81 m s$^{-2}$

Relative roughness of the pipeline = $6.00 \times 10^{-4}$

7   Water to be used for sluicing is led from a reservoir through a pipe (inside diameter = 15.0 cm, length = 584 m), to a pump, then through another pipe (inside diameter = 12.0 cm, length = 276 m) before discharging through a short nozzle (outlet diameter = 3.50 cm). If the level of water in the reservoir is 57.3 m above the discharge nozzle, and the pump is rated at 5.75 kW, find:
(i) the velocity of water emerging from the nozzle,
(ii) the volumetric flow rate of water through the nozzle, and
(iii) the power of the water jet.

Density of water = 999 kg m$^{-3}$

Acceleration due to gravity = 9.81 m s$^{-2}$

Friction factor, $\phi$, for both pipelines = 0.0035

8   Sulphuric acid is to be moved from a storage tank, through a pipeline (inside diameter = 5.50 cm, total length = 118 m), to fill rail cars. The free surface of acid in the storage tank is 18.6 m above the point at which liquid will discharge from the pipeline. If there are fittings in the pipeline which have a total resistance equivalent to 470 pipe diameters of straight pipe, calculate the power of pump required to move the acid at 12.0 dm$^3$ s$^{-1}$.

Density of the acid = 1846 kg m$^{-3}$

Viscosity of the acid = $3.21 \times 10^{-2}$ kg m$^{-1}$ s$^{-1}$

Relative roughness of the pipe = 0.0080

Acceleration due to gravity = 9.81 m s$^{-2}$

9  A formic acid solution (density = 1218 kg m$^{-3}$, viscosity = 9.17 × 10$^{-4}$ kg m$^{-1}$ s$^{-1}$) is being run into carboys from a storage tank through a length of smooth hose (inside diameter = 3.00 cm, length = 38.5 m). Calculate the volumetric flow rate of the solution when it is allowed to drain freely through the hose, and the free surface of liquid in the tank is 9.25 m above the hose outlet. Changes in kinetic energy may be neglected.

Acceleration due to gravity = 9.81 m s$^{-2}$

10  Water is to be drawn from a canal, through a smooth pipeline (length = 117 m) which will discharge freely at a point 8.25 m below the surface of water in the canal. A flow rate of 4.00 dm$^3$ s$^{-1}$ is required, and this must be obtained without pumping. Find the minimum diameter of pipe required. Changes in kinetic energy of the fluid may be neglected.

Density of water = 999 kg m$^{-3}$

Viscosity of water = 1.22 × 10$^{-3}$ kg m$^{-1}$ s$^{-1}$

Acceleration due to gravity = 9.81 m s$^{-2}$

For smooth pipes, $\phi = 0.0396\ Re^{-0.25}$

## 1.3 References

References are abbreviated as follows:

C & R (1) = J. M. Coulson & J. F. Richardson, Chemical Engineering, Vol. 1, 3rd edition, Pergamon, 1977

McC, S & H = W. L. McCabe, J. C. Smith and P. Harriott, Unit Operations of Chemical Engineering, 4th edition, McGraw-Hill, 1985

D, G & S = J. F. Douglas, J. M. Gasiorek & J. A. Swaffield, Fluid Mechanics, Pitman, 1980

1. C & R (1), page 9
2. C & R (1), page 39
3. McC, S & H, page 43
4. D, G & S, page 106
5. C & R (1), page 52
6. McC, S & H, page 78
7. D, G & S, page 251
8. C & R (1), page 100
9. McC, S & H, page 29
10. D, G & S, page 44
11. McC, S & H, page 77
12. D, G & S, pages 250, 251
13. C & R (1), page 51
14. McC, S & H, page 76
15. C & R (1), page 315
16. D, G & S, page 291
17. C & R (1), page 56
18. McC, S & H, page 86
19. C & R (1), page 27
20. C & R (1), page 43
21. McC, S & H, page 68
22. D, G & S, page 160
23. C & R (1), page 57
24. McC, S & H, page 85
25. D, G & S, page 165

# 1.4 Notation

| Symbol | Description | Unit |
|---|---|---|
| $d$ | diameter of pipe | m |
| $d_p$ | diameter of a drain tube | m |
| $d_t$ | diameter of a tank | m |
| $D$ | depth of liquid in a tank | m |
| $e$ | absolute roughness of a pipe wall | m |
| $F$ | energy loss due to friction for unit mass of liquid | $J\,kg^{-1}$ |
| $g$ | acceleration due to gravity | $m\,s^{-2}$ |
| $G$ | mass flow rate | $kg\,s^{-1}$ |
| $h_1$ | height of a column of liquid at section 1 | m |
| $h_2$ | height of a column of liquid at section 2 | m |
| $l$ | length of pipe | m |
| $l_p$ | length of a drain tube | m |
| $P$ | pressure in liquid | Pa |
| $P_1$ | pressure in liquid at section 1 | Pa |
| $P_2$ | pressure in liquid at section 2 | Pa |
| $P_{atm}$ | atmospheric pressure | Pa |
| $P'_2$ | gauge pressure in a liquid at section 2 | Pa |
| $Q$ | volumetric flow rate of a liquid | $m^3\,s^{-1}$ |
| $r$ | radius of a pipe | m |
| $R$ | shear stress experienced at the pipe wall | $N\,m^{-2}$ |
| $R_p$ | shear stress experienced at the wall of a drain tube | $N\,m^{-2}$ |
| $Re$ | pipe-flow Reynolds number | |
| $Re_p$ | Reynolds number for liquid flowing in a drain tube | |
| $s$ | distance from pipe axis | m |
| $t$ | time | s |
| $u$ | average velocity of liquid | $m\,s^{-1}$ |
| $u_1$ | average velocity of liquid in a tank | $m\,s^{-1}$ |
| $u_2$ | average velocity of liquid in a drain tube | $m\,s^{-1}$ |
| $\dot{u}_s$ | velocity of liquid along the axis of a pipe | $m\,s^{-1}$ |
| $\dot{u}_x$ | velocity of liquid at a distance $s$ from the axis of a pipe | $m\,s^{-1}$ |
| $W_s$ | work done by unit mass of liquid | $J\,kg^{-1}$ |
| $x$ | $(0.52 + D)$, (see p. 19) | m |
| $y$ | distance from pipe wall | m |
| $\alpha$ | kinetic energy correction factor | — |
| $\Delta h$ | $h_2 - h_1$. | m |
| $\Delta P$ | pressure difference between sections 2 and 1 | Pa |
| $\Delta P_p$ | pressure difference across a drain tube | Pa |
| $\Delta u^2$ | change in the square of average liquid velocity | $m^2\,s^{-2}$ |
| $\Delta z$ | difference in liquid levels | m |
| $\mu$ | viscosity of liquid | $kg\,m^{-1}\,s^{-1}$ |
| $\rho$ | density of liquid | $kg\,m^{-3}$ |
| $\phi$ | Stanton–Pannell friction factor | |
| $\phi_p$ | friction factor for liquid moving in a drain tube | |
| $\phi_t$ | friction factor for liquid moving in a tank | |

# 2 Pipe flow of gases and gas–liquid mixtures

## 2.1 Worked Examples

### 2.1.1 Gas flow through a pipeline when compressibility must be considered

*Problem*

What pressure must be maintained by a compressor at the entrance to a horizontal pipe (68.5 m long, 8.4 cm I.D.) in order to pump methane isothermally at 24 °C, at a rate of 2.35 kg s$^{-1}$, to a reactor in which the pressure is to be 560 kPa?

Molar mass of methane = $16.04 \times 10^{-3}$ kg mol$^{-1}$
Molar volume of a gas = $22.414 \times 10^{-3}$ m$^3$ mol$^{-1}$ at S.T.P.
Universal gas constant = 8.314 J mol$^{-1}$ K$^{-1}$
Viscosity of methane = $1.08 \times 10^{-5}$ kg m$^{-1}$ s$^{-1}$
Pipe roughness = 0.026 cm
1 atmosphere pressure = 101.3 kPa

*Solution*

The mechanical energy balance equation may be modified to allow for compressibility and flow on a mass basis. The result may be written:

$$\left(\frac{G}{A}\right)^2 \ln\frac{P_1}{P_2} + \frac{P_2^2 - P_1^2}{2RT/M} + 4\phi\frac{l}{d}\left(\frac{G}{A}\right)^2 = 0 \qquad (2.1)$$

if the pipe through which gas is flowing is horizontal and flow is turbulent (ref. 1, 2 and 3).

For this equation, cross-sectional area of the pipe,

$$A = \pi(4.2 \times 10^{-2})^2 \text{ m}^2$$
$$= 5.54 \times 10^{-3} \text{ m}^2$$

Friction factor, $\phi$, may be evaluated from a $\phi$ versus $Re$ chart if the Reynolds number for flow and relative roughness of the pipe are calculated. Since curves on this chart are almost parallel to the Reynolds number axis at higher Reynolds numbers, it will be sufficient to calculate $Re$ for conditions at the reactor end of the pipe where gas pressure, and therefore gas density and velocity, can be calculated readily.

At S.T.P., $\qquad$ gas density $= \dfrac{16.04 \times 10^{-3}}{22.414 \times 10^{-3}}$ kg m$^{-3}$

At 560 kPa and 24 °C, gas density,

$$\rho = \dfrac{16.04 \times 10^{-3} \times 560 \times 273}{22.414 \times 10^{-3} \times 101.3 \times 297} \text{ kg m}^{-3}$$

$$= 3.636 \text{ kg m}^{-3}$$

Thus, volumetric flow rate at the reactor entrance

$$= \dfrac{2.35}{3.636} \text{ m}^3 \text{ s}^{-1}$$

and average fluid velocity there,

$$u = \dfrac{2.35}{3.636 \times 5.54 \times 10^{-3}} \text{ m s}^{-1}$$

$$= 116.7 \text{ m s}^{-1}$$

Now, Reynolds number for fluid flow

$$= \dfrac{3.636 \times 116.7 \times 8.4 \times 10^{-2}}{1.08 \times 10^{-5}}$$

$$= 3.30 \times 10^6$$

so flow will be turbulent.

Furthermore, relative roughness of the pipe $= \dfrac{0.026}{8.4} = 0.003\,10$

So from a $\phi$ versus $Re$ chart, $\phi = 0.0032$
Now, substituting data into equation (2.1),

$$\dfrac{2.35^2}{(5.54 \times 10^{-3})^2} \ln \dfrac{P_1}{560 \times 10^3} + \dfrac{(560 \times 10^3)^2 - P_1^2}{(2 \times 8.314 \times 297)/(16.04 \times 10^{-3})}$$

$$+ \dfrac{4 \times 0.0032 \times 68.5 \times 2.35^2}{8.4 \times 10^{-2} \times (5.54 \times 10^{-3})^2} = 0$$

from which, $\qquad \ln P_1 - 1.805 \times 10^{-11} P_1^2 + 2.863 = 0$

This equation may be solved graphically, or by trial and error.

Hence, $\qquad P_1 = 0.057$ Pa or 960.1 kPa

Since, in a pipe of uniform cross section, a negative pressure gradient must be established to cause flow, the pressure to be maintained by the compressor at the pipe entrance = 960.1 kPa.

## 2.1.2 Flow of an ideal gas at maximum velocity under isothermal and adiabatic conditions

*Problem*

Find (i) the critical pressure ratio, $P_w/P_1$, (ii) the maximum velocity and (iii) the maximum mass flow rate at which air could be admitted in isothermal flow to a chamber, from atmosphere, through a horizontal pipe (34 cm long, 4 mm I.D.), if pressure in the chamber can be maintained at 15 kPa. (iv) What will be the rate at which heat must be transferred through the pipe wall to maintain isothermal conditions?

What would be (v) the critical pressure ratio, $P_w/P_1$, (vi) the maximum velocity, and (vii) the maximum mass flow rate if flow and expansion were to take place adiabatically? (viii) What would be the temperature of the air as it emerged from the pipe after adiabatic expansion?

Molar mass of air = $28.8 \times 10^{-3}$ kg mol$^{-1}$
Viscosity of air = $1.73 \times 10^{-5}$ kg m$^{-1}$ s$^{-1}$
Heat capacity ratio, $\gamma = C_p/C_v$, for air = 1.402
Molar volume of a gas = $22.414 \times 10^{-3}$ m$^3$ mol$^{-1}$ at S.T.P.
Universal gas constant = 8.314 J mol$^{-1}$ K$^{-1}$
Atmospheric pressure on the day = 107.6 kPa
Temperature on the day = 12 °C
Pipe roughness = 0.0032 mm

*Solution*

(i) Equation (2.1) which describes isothermal flow may be derived from the overall energy balance equation. Recognising that under maximum flow conditions $(G_w/A)^2 = P_w/v_w$, this equation may be rearranged to give (ref. 4):

$$\ln\left(\frac{P_1}{P_w}\right)^2 + 1 - \left(\frac{P_1}{P_w}\right)^2 + 8\phi\frac{l}{d} = 0 \quad (2.2)$$

Critical pressure ratio ($P_w/P_1$) with turbulent isothermal flow in a horizontal pipe may be deduced from equation (2.2), in which $P_1$ is the pressure in the upstream reservoir and $P_w$ is the minimum pressure which can obtain in the pipe at its outfall end.

Since flow in the pipe may be expected to be turbulent, $\phi$ versus $Re$ curves lie almost parallel to the Reynolds number axis. Friction factor $\phi$ will therefore be almost constant, and an estimate of it will suffice in the first instance.

Thus, relative roughness of the pipe $= \dfrac{0.0032}{4} = 0.0008$

So, if as a first estimate, a Reynolds number of approximately $5 \times 10^4$ is anticipated, we may take $\phi = 0.00285$.

Inserting data into equation (2.2),

$$\ln\left(\frac{P_1}{P_w}\right)^2 + 1 - \left(\frac{P_1}{P_w}\right)^2 + \frac{8 \times 0.00285 \times 34 \times 10^{-2}}{4 \times 10^{-3}} = 0$$

so that
$$\ln\left(\frac{P_1}{P_w}\right)^2 - \left(\frac{P_1}{P_w}\right)^2 + 2.938 = 0$$

This equation may be solved graphically, or by trial and error.

Hence, $\qquad P_1/P_w = 0.236$ or $2.104$

Since $P_1$ must be greater than $P_w$, 2.104 must be the correct value for $P_1/P_w$. So critical pressure ratio,

$$P_w/P_1 = 1/2.104 = 0.475$$

(ii) Maximum fluid velocity in isothermal flow (ref. 5 and 6),

$$u_w = (P_w v_w)^{0.5}$$

Since $\qquad P_w/P_1 = 0.475$

$\qquad P_w = 0.475 \times 107.6 = 51.11\ \mathrm{kPa}$

Now the density of air $= \dfrac{28.8 \times 10^{-3}}{22.414 \times 10^{-3}}\ \mathrm{kg\,m^{-3}}$ at S.T.P.

At 51.11 kPa and 12 °C, air density,

$$\rho_w = \frac{28.8 \times 10^{-3} \times 51.11 \times 273}{22.414 \times 10^{-3} \times 107.6 \times 285}\ \mathrm{kg\,m^{-3}}$$

$$= 0.585\ \mathrm{kg\,m^{-3}}$$

So specific volume, $v_w = 1/0.585 = 1.709\ \mathrm{m^3\,kg^{-1}}$

and maximum velocity with which air will be discharged into the chamber,

$$u_w = (51.11 \times 10^3 \times 1.709)^{0.5}\ \mathrm{m\,s^{-1}}$$

$$= \underline{295.5\ \mathrm{m\,s^{-1}}}$$

(iii) The maximum mass flow rate at which air would enter the low-pressure chamber,

$$G_w = Au_w/v_w$$

But since $\quad u_w = (P_w v_w)^{0.5}$ (ref. 5 and 6)

$\quad v_w = u_w^2/P_w$

and $\quad G_w = AP_w/u_w$

Now substituting data, the maximum mass flow rate,

$$G_w = \frac{\pi(2 \times 10^{-3})^2 \times 51.11 \times 10^3}{295.5} \text{ kg s}^{-1}$$

$$= \underline{2.173 \text{ g s}^{-1}}$$

At this stage, the value taken for the Reynolds number should be checked. This will be done for flow in the effluent end of the pipe.

Here, $\quad Re = \dfrac{\rho_w u_w d}{\mu}$

But, $\quad u_w = G_w/(\rho_w A)$

$\quad = \dfrac{4G_w}{\pi \rho_w d^2}$

So, $\quad Re = \dfrac{4G_w}{\pi d \mu}$

$\quad = \dfrac{4 \times 2.173 \times 10^{-3}}{\pi \times 4 \times 10^{-3} \times 1.73 \times 10^{-5}}$

$\quad = 4.00 \times 10^4$

This is close to the $Re$ value anticipated, so the value adopted for friction factor $\phi$ was almost correct.

(iv) Under isothermal conditions, heat taken up by the system is used entirely to increase the kinetic energy of the gas (ref. 7). So the rate of uptake of heat necessary to maintain isothermal conditions

$$= G \cdot \Delta\left(\frac{u^2}{2}\right)$$

$$= 2.173 \times 10^{-3} \left(\frac{295.5^2 - 0^2}{2}\right) \text{ J s}^{-1}$$

$$= \underline{94.87 \text{ J s}^{-1}}$$

(v) Under adiabatic conditions, it is convenient to find first of all the specific volume, $v_2$, of air as it emerges from the pipe under the particular pipe and upstream conditions which obtain in the system. This may be done by using the equation (ref. 8):

$$8\phi\frac{l}{d} = \left(1 - \frac{v_1^2}{v_2^2}\right)\left[\frac{\gamma-1}{2\gamma} + \frac{P_1}{v_1}\left(\frac{A}{G}\right)^2\right] - \left(\frac{\gamma+1}{\gamma}\right)\ln\frac{v_2}{v_1} \quad (2.3)$$

Mass flow rate, $G$, however is not known explicitly at this stage. But an energy balance on the fluid which is flowing adiabatically and turbulently, in a horizontal pipe, when no work is being done on or by the fluid, can be shown (ref. 9) to give rise to:

$$\frac{1}{2}\left(\frac{G}{A}\right)^2 v^2 + \left(\frac{\gamma}{\gamma-1}\right)Pv = \text{constant}$$

$$= \frac{1}{2}\left(\frac{G}{A}\right)^2 v_1^2 + \left(\frac{\gamma}{\gamma-1}\right)P_1 v_1 \quad (2.4)$$

$$= \frac{1}{2}\left(\frac{G}{A}\right)^2 v_2^2 + \left(\frac{\gamma}{\gamma-1}\right)P_2 v_2 \quad (2.5)$$

From equation (2.4)–(2.5)

$$\left(\frac{G}{A}\right)^2 = \left(\frac{2\gamma}{\gamma-1}\right)\left(\frac{P_2 v_2 - P_1 v_1}{v_1^2 - v_2^2}\right) \quad (2.6)$$

Substituting equation (2.6) into equation (2.3) and rearranging,

$$\left(1 - \frac{v_1^2}{v_2^2}\right)\left(\frac{\gamma-1}{2\gamma}\right)\left[1 + \frac{P_1}{v_1}\left(\frac{v_1^2 - v_2^2}{P_2 v_2 - P_1 v_1}\right)\right] - \left(\frac{\gamma+1}{\gamma}\right)\ln\frac{v_2}{v_1} - 8\phi\frac{l}{d} = 0 \quad (2.7)$$

Also, under adiabatic conditions,

$$P_1 v_1^\gamma = P_2 v_2^\gamma = \text{constant}$$

so

$$P_2 = P_1\left(\frac{v_1}{v_2}\right)^\gamma \quad (2.8)$$

Substituting equation (2.8) into equation (2.7) and simplifying,

$$\left(1 - \frac{v_1^2}{v_2^2}\right)\left(\frac{\gamma-1}{2\gamma}\right)\left[1 + \frac{v_1^2 - v_2^2}{v_1 v_2\left(\frac{v_1}{v_2}\right)^\gamma - v_1^2}\right] - \left(\frac{\gamma+1}{\gamma}\right)\ln\frac{v_2}{v_1} - 8\phi\frac{l}{d} = 0 \quad (2.9)$$

Values can be put to all parameters except $v_2$ in this equation, so it can be used to solve for $v_2$. Because friction factor $\phi$ is almost constant, its value may continue to be taken as:

$$\phi = 0.00285$$

Furthermore, $v_1 = \dfrac{RT_1}{MP_1}$

$$= \dfrac{8.314 \times 285}{28.8 \times 10^{-3} \times 107.6 \times 10^3} = 0.765 \text{ m}^3\text{kg}^{-1}$$

Substituting data into equation (2.9),

$$\left(1 - \dfrac{0.765^2}{v_2^2}\right)\left(\dfrac{1.402-1}{2 \times 1.402}\right)\left(1 + \dfrac{0.765^2 - v_2^2}{0.765^{1+\gamma} v_2^{1-\gamma} - 0.765^2}\right)$$
$$- \left(\dfrac{1.402+1}{1.402}\right)\ln\dfrac{v_2}{0.765} - \dfrac{8 \times 0.00285 \times 34 \times 10^{-2}}{4 \times 10^{-3}} = 0$$

so that

$$\left(0.1434 - \dfrac{0.0839}{v_2^2}\right)\left(1 + \dfrac{0.5852 - v_2^2}{0.5255 v_2^{-0.402} - 0.5852}\right) - 1.713 \ln \dfrac{v_2}{0.765} - 1.938 = 0$$
(2.10)

Equation (2.10) may be solved graphically.

Hence, $\quad\quad\quad\quad v_2 = \underline{2.79 \text{ m}^3 \text{kg}^{-1}}$

Now, because expansion is adiabatic,

$$P_1 v_1^\gamma = P_2 v_2^\gamma$$

so
$$P_2 = P_1 \left(\dfrac{v_1}{v_2}\right)^\gamma$$

and
$$P_2 = 107.6 \times 10^3 \left(\dfrac{0.765}{2.79}\right)^{1.402} \text{ Pa}$$

$$= \underline{17.54 \text{ kPa}}$$

The critical pressure ratio which would obtain under adiabatic conditions, is therefore,

$$\dfrac{P_2}{P_1} = \dfrac{P_w}{P_1} = \dfrac{17.54}{107.6}$$

$$= \underline{0.163}$$

(vi) The velocity with which air would emerge from the pipe is then, (ref. 10, 11 and 12),

$$u_w = (\gamma P_2 v_2)^{0.5}$$
$$= (1.402 \times 17.54 \times 10^3 \times 2.79)^{0.5} \text{ m s}^{-1}$$
$$= \underline{261.9 \text{ m s}^{-1}}$$

This would be the maximum (sonic) velocity.

## 34  Problems in fluid flow

(vii) The mass flow rate which would obtain under these conditions,

$$G_w = \frac{A u_w}{v_2}$$

$$= \frac{\pi (4 \times 10^{-3})^2 \times 261.9}{4 \times 2.79} \, \text{kg s}^{-1}$$

$$= \underline{1.180 \, \text{g s}^{-1}}$$

(viii) To find the temperature of the gas discharging, the result of manipulating the energy-balance equation [see equations (2.4) and (2.5)] may be invoked, i.e.:

$$\left(\frac{\gamma}{\gamma - 1}\right) Pv + \frac{1}{2}\left(\frac{G}{A}\right)^2 v^2 = \text{constant}$$

Since

$$Pv = \frac{RT}{M}$$

one may also write:

$$\text{constant} = \left(\frac{\gamma}{\gamma - 1}\right) P_1 v_1 + \frac{1}{2}\left(\frac{G}{A}\right)^2 v_1^2 \tag{2.11}$$

which also

$$= \left(\frac{\gamma}{\gamma - 1}\right)\frac{RT_2}{M} + \frac{1}{2}\left(\frac{G}{A}\right)^2 \left(\frac{RT_2}{MP_2}\right)^2 \tag{2.12}$$

Inserting data into equations (2.11) and (2.12),

$$\left(\frac{1.402}{1.402 - 1}\right)\frac{8.314 T_2}{28.8 \times 10^{-3}} + \frac{1}{2}\left(\frac{1.180 \times 10^{-3}}{\pi (2 \times 10^{-3})^2}\right)^2 \left(\frac{8.314 T_2}{28.8 \times 10^{-3} \times 17.54 \times 10^3}\right)^2$$

$$= \left(\frac{1.402}{1.402 - 1}\right) 107.6 \times 10^3 \times 0.765 + \frac{1}{2}\left(\frac{1.180 \times 10^{-3}}{\pi (2 \times 10^{-3})^2}\right)^2 0.765^2$$

Simplifying,

$$1.194 T_2^2 + 1006.8 T_2 - 2.897 \times 10^5 = 0$$

so that

$$T_2 = -1070 \, \text{K} \ \text{or} \ 226.8 \, \text{K}$$

Since the first of these figures represents an impossibility, the temperature at which air would emerge from the pipe,

$$T_2 = \underline{-46.2 \, ^\circ\text{C}}$$

## 2.1.3 Flow of a non-ideal gas at maximum velocity under adiabatic conditions

*Problem*

If steam, held in a chamber at a pressure of 550 kPa and temperature 350 °C, is discharged adiabatically through a pipe (length = 0.85 m, I.D. = 2.4 cm, absolute roughness = 0.096 mm) to atmosphere (pressure = 100.4 kPa), what will be (i) the mass flow rate of steam through the pipe, (ii) the pressure in the pipe at the downstream end of the pipe, and (iii) the temperature of steam emerging from the pipe? (iv) Establish whether a greater mass flow rate of steam would be obtained by reducing pressure around the discharge end of the pipe.

Ratio of heat capacities, $C_p/C_v$, for steam = 1.33
Enthalpy of steam at 550 kPa and 350 °C = 3167 kJ kg$^{-1}$
Molar mass of steam = $18.00 \times 10^{-3}$ kg mol$^{-1}$
Universal gas constant = 8.314 J mol$^{-1}$ K$^{-1}$

*Solution*

Steam at high pressure is a non-ideal fluid, so allowance must be made for its non ideality by using enthalpy data measured experimentally rather than using equations for ideal, adiabatic flow.

An energy balance for flow of fluid in an infinitesimally small length of pipe may be written:

$$dU + d(Pv) + d(u^2/2\alpha) + g\,dz = \delta q - \delta W_s \quad (2.13)$$

($\delta q$ is the net quantity of heat taken up by the fluid.) Thus, if one considers adiabatic, turbulent flow, in a horizontal pipe with no work being done on or by the fluid, and use the definition of enthalpy, equation (2.13) reduces to:

$$dH + u \cdot du = 0 \quad (2.14)$$

Now mass flow rate, $\quad G = Au\rho = Au/v$

So, $$u = \frac{G}{A} \cdot v \quad (2.15)$$

and $$du = \frac{G}{A} \cdot dv \quad (2.16)$$

Substituting equations (2.15) and (2.16) into equation (2.14),

$$dH + \left(\frac{G}{A}\right)^2 v \cdot dv = 0$$

so that on integrating,

$$H + \frac{1}{2}\left(\frac{G}{A}\right)^2 v^2 = K \text{ (a constant)} \quad (2.17)$$

36   Problems in fluid flow

$K$ may be evaluated by introducing data which apply to any particular location. Now for compressible flow, (ref. 13),

$$\left(\frac{G}{A}\right)^2 \ln\left(\frac{v_2}{v_1}\right) + \int_{P_1}^{P_2} \frac{dP}{v} + 4\phi\frac{l}{d}\left(\frac{G}{A}\right)^2 = 0 \tag{2.18}$$

To develop this further, equations (2.4) and (2.5) may be rearranged to obtain an expression for $P$ for adiabatic flow. This may then be differentiated to obtain an expression for $dP$, and the result used to evaluate $\int_{P_1}^{P_2} dP/v$. In this way, it may be shown that for adiabatic flow,

$$\int_{P_1}^{P_2} \frac{dP}{v} = \frac{\gamma-1}{\gamma}\left[\frac{K}{2}\left(\frac{1}{v_2^2}-\frac{1}{v_1^2}\right) - \frac{1}{2}\left(\frac{G}{A}\right)^2 \ln\frac{v_2}{v_1}\right] \tag{2.19}$$

So using equation (2.17), particularised for conditions at section 1 (i.e. in the steam chamber upstream), to substitute for $K$ in equation (2.19), then substituting the result into equation (2.18)

$$\left(\frac{G}{A}\right)^2 \ln\left(\frac{v_2}{v_1}\right) + \frac{\gamma-1}{\gamma}\left[\frac{H_1}{2}\left(\frac{1}{v_2^2}-\frac{1}{v_1^2}\right) + \frac{1}{4}\left(\frac{G}{A}\right)^2 v_1^2\left(\frac{1}{v_2^2}-\frac{1}{v_1^2}\right) - \frac{1}{2}\left(\frac{G}{A}\right)^2 \ln\left(\frac{v_2}{v_1}\right)\right]$$

$$+ 4\phi\frac{l}{d}\left(\frac{G}{A}\right)^2 = 0$$

On simplifying,

$$\ln\left(\frac{v_2}{v_1}\right) + \frac{\gamma-1}{\gamma}\left[\frac{H_1}{2}\left(\frac{A}{G}\right)^2\left(\frac{1}{v_2^2}-\frac{1}{v_1^2}\right) + \frac{1}{4}\left(\frac{v_1^2}{v_2^2}-1\right) - \frac{1}{2}\ln\left(\frac{v_2}{v_1}\right)\right] + 4\phi\frac{l}{d} = 0 \tag{2.20}$$

Now we wish to calculate $G$ for a value of $v_2$ which is as yet unknown. However, the atmosphere into which the steam is discharging has a pressure,

$$P_2 = \frac{100.4}{550} P_1 = 0.183 P_1$$

In a short nozzle, however, gas attains sonic velocity in the throat when $P_2 \approx 0.5 P_1$, when maximum flow rate will obtain through the system. In this case, the discharge pipe is 0.85 m in length. Thus, one may estimate that sonic velocity, and hence maximum gas flow rate, would obtain when pressure in the downstream end of the pipe, $P_w \approx 0.4 P_1$. Thus one may expect that in the system under consideration, steam would be discharging at a maximum rate. This maximum, $G_w$, would obtain when $dG/dv_2 = 0$, and then $v_2 = v_w$. So, rearranging equation (2.20),

$$\ln\left(\frac{v_2}{v_1}\right) + \frac{\gamma-1}{\gamma}\left[\frac{H_1}{2}A^2G^{-2}v_2^{-2} - \frac{H_1A^2G^{-2}}{2v_1^2} + \frac{v_1^2v_2^{-2}}{4} - \frac{1}{4} - \frac{1}{2}\ln\left(\frac{v_2}{v_1}\right)\right]$$

$$+ 4\phi\frac{l}{d} = 0$$

Differentiating this with respect to $v_2$,

$$\frac{v_1}{v_2}\cdot\frac{1}{v_1}+\frac{\gamma-1}{\gamma}\left[-H_1A^2G^{-3}v_2^{-2}\cdot\frac{dG}{dv_2}-H_1A^2G^{-2}v_2^{-3}\right.$$

$$\left.+\frac{H_1A^2}{v_1^2}\cdot G^{-3}\cdot\frac{dG}{dv_2}-\frac{v_1^2v_2^{-3}}{2}-\frac{1}{2}\cdot\frac{v_1}{v_2}\cdot\frac{1}{v_1}\right]=0$$

Setting $dG/dv_2$ to 0 and simplifying,

$$\frac{1}{v_w}-\frac{\gamma-1}{\gamma}\left(\frac{H_1A^2}{G_w^2v_w^3}+\frac{v_1^2}{2v_w^3}+\frac{1}{2v_w}\right)=0$$

from which

$$\frac{H_1A^2}{G_w^2}=\frac{\gamma v_w^2}{\gamma-1}-\frac{v_1^2}{2}-\frac{v_w^2}{2} \tag{2.21}$$

Now a value for $v_w$ is required in order to calculate $G_w$. $v_w$ can only be estimated, however, but the correct selection will yield a value of $G_w$ from equation (2.21) which on substitution back into equation (2.20) will yield a value of $v_2 = v_w$. To this end, for adiabatic expansion,

$$P_1v_1^\gamma = P_wv_w^\gamma$$

so

$$v_w = \left(\frac{P_1}{P_w}\right)^{1/\gamma}\cdot v_1$$

But since

$$P_w \approx 0.4 P_1$$

$$v_w \approx 2.5^{\frac{1}{1.33}} v_1$$

$$\approx 1.99 v_1$$

But

$$P_1v_1 = \frac{RT_1}{M}$$

so

$$v_1 = \frac{8.314 \times 623}{550 \times 10^3 \times 18.00 \times 10^{-3}}\text{ m}^3\text{ kg}^{-1}$$

$$= 0.5232 \text{ m}^3\text{ kg}^{-1}$$

Thus,

$$v_w \approx 1.99 \times 0.5232 \text{ m}^3\text{ kg}^{-1}$$

$$\approx 1.041 \text{ m}^3\text{ kg}^{-1}$$

Finally, the cross-sectional area of the pipe,

$$A = \frac{\pi(2.4 \times 10^{-2})^2}{4}\text{ m}^2$$

$$= 4.524 \times 10^{-4}\text{ m}^2$$

Now substituting data into equation (2.21),

$$\frac{3167 \times 10^3(4.524 \times 10^{-4})^2}{G_w^2}=\frac{1.33 \times 1.041^2}{0.33}-\frac{0.5232^2}{2}-\frac{1.041^2}{2}$$

## 38  Problems in fluid flow

from which, $\qquad G_w = 0.4191 \text{ kg s}^{-1}$

The value of $v_2$ which is required to make $G_w = 0.4191 \text{ kg s}^{-1}$ can now be obtained from equation (2.20). Firstly, however, we need a value for friction factor $\phi$.

Now relative roughness of the pipe

$$= \frac{0.096 \times 10^{-3}}{2.4 \times 10^{-2}} = 0.0040$$

Since a high value of Reynolds number may be expected, from a chart of $\phi$ versus $Re$,

$$\phi = 0.0035$$

Now substituting data into equation (2.20)

$$\ln v_2 - \ln 0.5232 + \frac{0.33}{1.33}\left[\frac{3167 \times 10^3}{2}\left(\frac{4.524 \times 10^{-4}}{0.4191}\right)^2\left(\frac{1}{v_2^2} - \frac{1}{0.5232^2}\right)\right.$$
$$\left. + \frac{1}{4}\left(\frac{0.5232^2}{v_2^2} - 1\right) - \frac{1}{2}\ln v_2 + \frac{1}{2}\ln 0.5232 \right] + \frac{4 \times 0.0035 \times 0.85}{2.4 \times 10^{-2}} = 0$$

Simplifying,

$$\ln v_2 + \frac{0.5421}{v_2^2} - 0.7662 = 0$$

This equation may be solved graphically, whence:

$$v_2 = 1.82 \text{ m}^3 \text{kg}^{-1}$$

This compares with the original estimate,

$$v_w = 1.041 \text{ m}^3 \text{kg}^{-1}$$

So the estimate of $v_w$ must be improved. If $P_w$ is considered $= 0.45 P_1$ it may be shown as above that $v_w = 0.954 \text{ m}^3 \text{kg}^{-1}$

$$G_w = 0.4591 \text{ kg s}^{-1}$$

and $\qquad v_2 = 0.954 \text{ m}^3 \text{kg}^{-1}$

Thus $v_2 = v_w$, so we may consider:
(i) mass flow rate of steam, $G_w$, through the pipe

$$= \underline{0.4591 \text{ kg s}^{-1}}$$

(ii) pressure in the pipe at the downstream end of the pipe

$$= 0.45 P_1$$
$$= 0.45 \times 550 \text{ kPa}$$
$$= \underline{247.5 \text{ kPa}}$$

(iii) To find the temperature, $T_w$, of steam as it emerges from the pipe, we may use equation (2.17) again,

$$H + \frac{1}{2}\left(\frac{G}{A}\right)^2 v^2 = \text{constant} \tag{2.17}$$

in conjunction with the approximation,

$$Pv = \frac{RT}{M}$$

Thus we may write: $\text{constant} = H_1 + \frac{1}{2}\left(\frac{G}{A}\right)^2 v_1^2 \tag{2.22}$

which also $= H_w + \frac{1}{2}\left(\frac{G}{A}\right)^2 \left(\frac{RT_w}{MP_w}\right)^2 \tag{2.23}$

Now it has been established that the pressure, $P_w$, at which steam will emerge = 247.5 kPa. Furthermore, $H_w$ will have a value which depends upon pressure and temperature. It is now necessary to select a temperature at which to evaluate, from tables of the thermodynamic properties of steam, $H_w$ which, when substituted into equation (2.22)–(2.23), will yield the same figure for $T_w$ at which $H_w$ was evaluated. Trial selections are necessary. If a temperature of 210 °C is selected at which to evaluate $H_w$, from steam tables it is found that $H_w = 2888.7 \text{ kJ kg}^{-1}$. Then substituting data into equations (2.22) and (2.23),

$$3167 \times 10^3 + \frac{0.4591^2 \times 0.5232^2}{2(4.524 \times 10^{-4})^2}$$

$$= 2888.7 \times 10^3 + \frac{0.4591^2 \times 8.314^2 \times T_w^2}{2(4.524 \times 10^{-4})^2 (18.00 \times 10^{-3})^2 (247.5 \times 10^3)^2}$$

from which, $T_w = 483 \text{ K}$
$= 210 \,°\text{C}$

So steam will emerge from the pipe at 210 °C.

(iv) Since maximum flow conditions obtain in the pipe, with sonic gas velocity at the exit, further expansion of the steam to a pressure of 100.4 kPa must take place after steam has reached the atmosphere, and further pressure reduction here would not increase mass flow rate.

### 2.1.4 Venting of gas from a pressure vessel

*Problem*
Provision is to be made to vent ethylene in an emergency from a reactor to the atmosphere through a short, horizontal, converging–diverging nozzle. The

## 40  Problems in fluid flow

reactor holds ethylene at 50 °C, and is required to discharge if pressure in the reactor reaches 3 atmospheres. The initial rate of discharge required is 18 kg s$^{-1}$. Calculate:

(i) the pressure which will obtain in the nozzle throat when 18 kg s$^{-1}$ is the maximum flow rate of ethylene able to discharge,

(ii) the diameter required for the nozzle throat, and

(iii) the velocity of gas in the throat when 18 kg of ethylene is discharging per second.

Molar mass of ethylene = $28.05 \times 10^{-3}$ kg mol$^{-1}$
Heat capacity ratio, $C_p/C_v$, for ethylene = 1.23
Universal gas constant = 8.314 J mol$^{-1}$ K$^{-1}$
1 atmosphere pressure = 101.3 kPa.

*Solution*

(i) Discharge will be rapid, so it may be considered to take place adiabatically. Thus, the pressure ratio for minimum duct area,

$$\frac{P_2}{P_1} = \left(\frac{2}{\gamma+1}\right)^{\gamma/(\gamma-1)} \quad \text{(ref. 14, 15 and 16)}$$

$$= \left(\frac{2}{2.23}\right)^{1.23/0.23}$$

$$= 0.5587$$

So the pressure, $P_2$, which must obtain in the nozzle throat

$$= 0.5587 \times 3 \times 101.3 \times 10^3 \text{ Pa}$$
$$= \underline{169.8 \text{ kPa}}$$

This is greater than atmospheric pressure, so maximum flow conditions will obtain in the nozzle.

(ii) The cross-sectional area required for the nozzle throat,

$$A_2 = \frac{Gv_2}{u_2} \quad (2.24)$$

But with adiabatic flow (ref. 14, 15 and 17),

$$u_2^2 = \frac{2\gamma P_1 v_1}{\gamma - 1}\left[1 - \left(\frac{P_2}{P_1}\right)^{(\gamma-1)/\gamma}\right] \quad (2.25)$$

Also,  $\quad P_1 v_1^\gamma = P_2 v_2^\gamma$

so that  $\quad v_2 = v_1 (P_2/P_1)^{-1/\gamma} \quad (2.26)$

Combining equations (2.24), (2.25) and (2.26)

$$A_2^2 = G^2 v_1^2 (P_2/P_1)^{-2/\gamma} \cdot \frac{\gamma - 1}{2\gamma P_1 v_1 [1 - (P_2/P_1)^{(\gamma-1)/\gamma}]} \tag{2.27}$$

Now, $\quad P_1 v_1 = RT_1/M$

so that $\quad v_1 = \dfrac{8.314 \times 323}{3 \times 101.3 \times 10^3 \times 28.05 \times 10^{-3}}\, \text{m}^3\,\text{kg}^{-1}$

$\qquad\qquad = 0.3150\, \text{m}^3\,\text{kg}^{-1}$

Now substituting data into equation (2.27),

$$A_2^2 = \frac{18^2 \times 0.3150 \times 0.5587^{-2/1.23} \times 0.23}{2 \times 1.23 \times 3 \times 101.3 \times 10^3 (1 - 0.5587^{0.23/1.23})}\, \text{m}^4$$

from which, $A_2 = 0.0280\, \text{m}^2$

So the diameter required for the nozzle throat

$$= \left(\frac{4 \times 0.0280}{\pi}\right)^{0.5}\, \text{m}$$

$$= \underline{18.88\, \text{cm}}$$

(iii) The velocity of fluid which will obtain in the throat at maximum rate of discharge is now given by equation (2.25). Thus,

$$u_2^2 = \frac{2 \times 1.23 \times 3 \times 101.3 \times 10^3 \times 0.3150}{0.23} (1 - 0.5587^{0.23/1.23})\, \text{m}^2\,\text{s}^{-2}$$

from which, gas velocity, $\quad u_2 = 325.0\, \text{m s}^{-1}$

This equals sonic velocity in the throat.

## 2.1.5 Gas-flow measurement with a venturi meter

*Problem*

(i) Derive an equation by which to express mass flow rate of a turbulently flowing gas stream in terms of pressure drop it induces in a short, horizontal, venturi meter tube and pressure in the throat when gas flows isothermally, and throat area is small compared with cross-sectional area of the pipeline into which the venturi tube is set.

(ii) What will be the pressure drop (measured in cm of water) observed across a venturi meter (throat diameter = 4.5 mm) when methane flows through the tube at 0.75 g s$^{-1}$, and pressure which obtains in the throat of the venturi tube is found to be 764.3 mm Hg?

## 42 Problems in fluid flow

Temperature of the day = 15 °C
Density of water at 15 °C = 999 kg m$^{-3}$
Density of mercury at 15 °C = 13 559 kg m$^{-3}$
Molar mass of methane = 16.04 × 10$^{-3}$ kg mol$^{-1}$
Universal gas constant = 8.314 J mol$^{-1}$ K$^{-1}$
Acceleration due to gravity = 9.81 m s$^{-2}$

*Solution*

(i) The overall energy balance equation may be written (ref. 18, 19 and 20):

$$\Delta\left(\frac{u^2}{2\alpha}\right) + g\Delta z + \int_{P_1}^{P_2} v\,dP + W_s + F = 0 \tag{2.28}$$

When fluid flows turbulently throughout the system, $\alpha \approx 1$.
When the venturi meter tube is horizontal, $\Delta z = 0$.
If the tube is short, $F \approx 0$. Finally, no work will be done on or by the fluid as it traverses the venturi meter tube, so $W_s = 0$. Thus equation (2.28) becomes:

$$\Delta(u^2) = -2\int_{P_1}^{P_2} v\,dP \tag{2.29}$$

Since gas flows isothermally,

$$Pv = RT/M = \text{constant} = P_1 v_1 = P_2 v_2 \tag{2.30}$$

So,
$$v = \frac{P_2 v_2}{P}$$

and equation (2.29) becomes:

$$\Delta(u^2) = -2P_2 v_2 \int_{P_1}^{P_2} \frac{dP}{P}$$

$$= -2P_2 v_2 \cdot \ln(P_2/P_1)$$

Since the throat area is small compared with the cross-sectional area of the pipeline into which the venturi tube is set,

$$u_2 \gg u_1$$

so we may write:
$$u_2^2 = 2P_2 v_2 \cdot \ln(P_1/P_2) \tag{2.31}$$

Now where $A$ = orifice area,

mass flow rate of gas, $\quad G = A u_2 \rho_2 = A u_2 / v_2 \tag{2.32}$

Combining equations (2.31) and (2.32),

$$G = \frac{A}{v_2}[2P_2 v_2 \cdot \ln(P_1/P_2)]^{0.5}$$

$$= A\left(2\frac{P_2}{v_2} \cdot \ln(P_1/P_2)\right)^{0.5}$$

Now using equation (2.30),

$$G = A\left(\frac{2P_2^2 M}{RT} \cdot \ln(P_1/P_2)\right)^{0.5} \quad (2.33)$$

But,
$$\frac{P_1}{P_2} = 1 - \frac{P_2 - P_1}{P_2}$$

$$= 1 - \frac{\Delta P}{P_2}$$

So equation (2.33) becomes:

$$G = A\left[\frac{2P_2^2 M}{RT} \cdot \ln\left(1 - \frac{\Delta P}{P_2}\right)\right]^{0.5} \quad (2.34)$$

This is the equation required.

(ii) Now, $P_2 = 764.3$ mm Hg

$$= 764.3 \times 10^{-3} \times 13\,559 \times 9.81 \text{ Pa} = 101.7 \text{ kPa}$$

Rearranging equation (2.34)

$$\ln\left(1 - \frac{\Delta P}{P_2}\right) = \frac{RTG^2}{2P_2^2 MA^2}$$

Inserting data,

$$\ln\left(1 - \frac{\Delta P}{P_2}\right) = \frac{8.314 \times 288(0.75 \times 10^{-3})^2 \times 4^2}{2(101.7 \times 10^3)^2 \times 16.04 \times 10^{-3} \times \pi^2 (4.5 \times 10^{-3})^4}$$

$$= 1.605 \times 10^{-2}$$

Taking antilogs,

$$1 - \frac{\Delta P}{101.7 \times 10^3} = 1.0162$$

so, $\quad -\Delta P = 0.0162 \times 101.7 \times 10^3 = 1648$ Pa

Since a manometer containing water will be used to measure this pressure drop,

$$1648 = \Delta h \times 999 \times 9.81$$

from which $\quad \Delta h = 16.8$ cm

Thus, the pressure drop observed will be $= 16.8$ cm $H_2O$

## 2.1.6 Pressure drop required for flow of a gas–liquid mixture through a pipe

*Problem*

Steam and water at 133 °C and a mean pressure of 300 kPa flow together through a pipe (0.075 m I.D.) at mass flow rates of 0.05 and 1.5 kg s$^{-1}$ respectively. Calculate the pressure drop per metre length of pipe which must obtain to drive the mixture assuming that no heat is exchanged with the surroundings.

Density of water at 133 °C = 931 kg m$^{-3}$
Viscosity of water at 133 °C = 1.55 × 10$^{-4}$ kg m$^{-1}$ s$^{-1}$
Specific volume of steam at 133 °C and 300 kPa = 0.6057 m$^3$ kg$^{-1}$
Viscosity of steam at 133 °C = 1.38 × 10$^{-5}$ kg m$^{-1}$ s$^{-1}$
Relative roughness of the pipe = 1.5 × 10$^{-4}$

*Solution*

The Lockhart–Martinelli method (ref. 21, 22 and 23) will be used.

$$\text{Cross-sectional area of the pipe} = \frac{\pi(0.075)^2}{4} = 4.42 \times 10^{-3} \text{ m}^2$$

Now considering the steam phase, its density at 133 °C and 300 kPa

$$= 1/0.6057 = 1.651 \text{ kg m}^{-3}$$

So the velocity of steam in the pipe if steam were to flow alone

$$= \frac{0.05}{1.651 \times 4.42 \times 10^{-3}} \text{ m s}^{-1}$$

$$= 6.85 \text{ m s}^{-1}$$

Now

$$Re_G = \frac{1.651 \times 6.85 \times 0.075}{1.38 \times 10^{-5}}$$

$$= 6.15 \times 10^4$$

Using a $\phi$ versus $Re$ chart, $\quad \phi_G = 0.00245$

Now under steady state conditions, the force on fluid in a pipeline, which arises due to pressure drop, is balanced by frictional forces which arise at the pipe wall. Thus:

$$\frac{-\Delta P \pi d^2}{4} = R_o \pi l d$$

so that
$$-\Delta P = 4 R_o \frac{l}{d}$$

$$= 4\phi (lu^2 \rho / d) \qquad (2.35)$$

So using equation (2.35),

$$-\Delta P_G = \frac{4 \times 0.00245 \times 1 \times 6.85^2 \times 1.651}{0.075} \text{ Pa per metre length of pipe}$$

$$= 10.12 \text{ Pa per metre length of pipe}$$

Considering now the water phase, its average velocity if it were to flow alone

$$= \frac{1.5}{931 \times 4.42 \times 10^{-3}} \text{ m s}^{-1}$$

$$= 0.365 \text{ m s}^{-1}$$

So

$$Re_L = \frac{931 \times 0.365 \times 0.075}{1.55 \times 10^{-4}}$$

$$= 1.64 \times 10^5$$

Using a $\phi$ versus $Re$ chart, $\phi_L = 0.00205$

Now using equation (2.35),

$$-\Delta P_L = \frac{4 \times 0.00205 \times 1 \times 0.365^2 \times 931}{0.075} \text{ Pa per metre length of pipe}$$

$$= 13.56 \text{ Pa per metre length of pipe}$$

Now parameter
$$X = \left(\frac{\Delta P_L}{\Delta P_G}\right)^{0.5} \quad \text{(refs. 21, 22 and 23)}$$

$$= \left(\frac{13.56}{10.12}\right)^{0.5} = 1.158$$

So since both steam and water would be in turbulent motion if flowing alone, (ref. 21, 22 and 23),

$$\Phi_G = 5.0$$

So, pressure drop required to drive the steam–water mixture through the pipe,

$$-\Delta P_{TPF} = 5.0^2 \times 10.12 \text{ Pa per metre length of pipe}$$

$$= \underline{253 \text{ Pa per metre length of pipe}}$$

## 2.2 Student Exercises

1 (i) A compressor is required to drive acetylene gas at 1.85 kg s$^{-1}$ through a horizontal pipeline, 68.7 m long. The maximum pressure that may be developed by the compressor = 1.34 MPa, and gas pressure at the delivery end of the pipeline must be 470 kPa. If the system is to operate isothermally at 20 °C, find the inside diameter of a smooth pipe which will be required to construct the pipeline.

(ii) If a pipeline of twice this diameter were actually to be installed, to what pressure must the gas be raised at the compressor if delivery pressure is to remain the same?

Molar mass of acetylene = $26.02 \times 10^{-3}$ kg mol$^{-1}$
Viscosity of acetylene = $1.01 \times 10^{-5}$ kg m$^{-1}$ s$^{-1}$
Universal gas constant = 8.314 J mol$^{-1}$ K$^{-1}$

2  If a gas mixture was to discharge in isothermal flow at 17 °C from a tank in which pressure = 685 kPa, to atmosphere, through a pipe of inside diameter = 7.8 mm, and relative roughness = 0.0002, calculate:
(i) the maximum length of pipe from which gas could emerge without being choked,
(ii) the mass flow rate of gas which would obtain then,
(iii) the pressure which would obtain in the discharge end of the pipe if it were only 3 m long,
(iv) the mass flow rate of gas mixture which would obtain through a pipe 3 m in length, and
(v) the rate at which heat would have to transfer through the pipe wall in part (ii) to keep flow isothermal.

Molar mass of the gas mixture = 0.022 kg mol$^{-1}$
Atmospheric pressure on the day = 102.8 kPa
Universal gas constant = 8.314 J mol$^{-1}$ K$^{-1}$

3  Methane gas escapes from a large tank, in which the pressure = 765 kPa, and temperature = 23°C, to atmosphere through a horizontal pipe (length = 5.85 m, inside diameter = 40.0 mm, relative roughness = 0.0006). If expansion occurs adiabatically,
(i) find the pressure in the discharge end of the pipe, and so establish that critical (maximum flow rate) conditions would obtain in the system,
(ii) calculate the mass flow rate of methane leaving the pipe, and
(iii) find the temperature of the methane as it leaves the pipe.

Molar mass of methane = 0.016 kg mol$^{-1}$
Heat capacity ratio, $\gamma = C_p/C_v$, for methane = 1.313
Atmospheric pressure on the day = 100.3 kPa
Universal gas constant = 8.314 J mol$^{-1}$ K$^{-1}$

4  Emergency discharge of compressed air is to take place from a storage tank through a short converging–diverging nozzle if pressure in the tank reaches 2 MPa. If air in the tank is at 280 K, and an initial discharge rate of 12 kg s$^{-1}$ is required in an emergency, find:
(i) the pressure which will obtain in the nozzle throat as air discharges through a throat of minimum cross-sectional area at 12 kg s$^{-1}$, and

(ii) the minimum diameter required for the nozzle throat.

Molar mass of air = $0.0288$ kg mol$^{-1}$
Heat capacity ratio, $\gamma$, for air = $1.402$
Universal gas constant = $8.314$ J mol$^{-1}$ K$^{-1}$

5 (i) A short, horizontal venturi meter is to be used to measure flow rate of a gas stream under isothermal conditions. Obtain equations to calculate mass flow rate and volumetric flow rate just upstream of the meter tube in terms of pressure in the throat and pressure drop between the upstream location and the throat, if cross-sectional area of the throat of the venturi tube is small compared with cross-sectional area of the main pipe.
(ii) What will be (a) the mass flow rate, (b) upstream volumetric flow rate, and (c) average upstream gas velocity if a venturi tube, with throat diameter = 3.50 cm, produces a pressure drop of 18.7 cm of water as chlorine gas passes through it in isothermal flow at 16 °C?

Diameter of the main pipe = $28.0$ cm
Gas pressure in the venturi tube throat = $102.8$ kPa
Density of water = $999$ kg m$^{-3}$
Molar mass of chlorine gas = $70.9 \times 10^{-3}$ kg mol$^{-1}$
Universal gas constant = $8.314$ J mol$^{-1}$ K$^{-1}$
Acceleration due to gravity = $9.81$ m s$^{-2}$

6 Calculate the pressure drop required to drive a mixture of gaseous and liquid ammonia at $0.0097$ and $0.095$ kg s$^{-1}$ respectively through a smooth, horizontal pipe of 20 mm inside diameter and length = $1.18$ m. Temperature and mean pressure in the system are 20 °C and 857 kPa respectively.

Density of liquid ammonia = $804$ kg m$^{-3}$
Viscosity of liquid ammonia = $1.22 \times 10^{-4}$ kg m$^{-1}$ s$^{-1}$
Density of gaseous ammonia at 20 °C and 857 kPa = $6.693$ kg m$^{-3}$
Viscosity of gaseous ammonia = $1.01 \times 10^{-5}$ kg m$^{-1}$ s$^{-1}$

## 2.3 References

References are abbreviated as follows:

C & R (1) = J. M. Coulson and J. F. Richardson, Chemical Engineering, Vol. 1, 3rd edition, Pergamon, 1977

McC, S & H = W. L. McCabe, J. C. Smith and P. Harriott, Unit Operations of Chemical Engineering, 4th edition, McGraw-Hill, 1985

D, G & S = J. F. Douglas, J. M. Gasiorek and J. A. Swaffield, Fluid Mechanics, Pitman, 1980

1. C & R (1), page 73
2. McC, S & H, page 123
3. D, G & S, page 428
4. C & R (1), page 76
5. C & R (1), page 75
6. McC, S & H, page 122
7. C & R (1), page 76
8. C & R (1), page 78
9. C & R (1), page 80
10. C & R (1), page 81
11. McC, S & H, page 109
12. D, G & S, page 414
13. C & R (1), page 72
14. C & R (1), page 85
15. McC, S & H, page 115
16. D, G & S, page 413
17. D, G & S, page 414
18. C & R (1), page 27
19. McC, S & H, page 108
20. D, G & S, page 160
21. R. W. Lockhart and R. C. Martinelli, *Chem. Eng. Prog.*, 1949, **45**, 39–48
22. C & R (1), page 92
23. F. A. Holland, Fluid Flow for Chemical Engineers, Edward Arnold, 1973, pages 133–8

## 2.4 Notation

| Symbol | Description | Unit |
|---|---|---|
| $A$ | cross-sectional area of duct | $m^2$ |
| $A_2$ | cross-sectional area of duct at section 2 | $m^2$ |
| $C_p$ | heat capacity of fluid at constant pressure | $J\,mol^{-1}\,K^{-1}$ |
| $C_v$ | heat capacity of fluid at constant volume | $J\,mol^{-1}\,K^{-1}$ |
| $d$ | pipe diameter | m |
| $F$ | energy loss per unit mass of fluid due to friction | $J\,kg^{-1}$ |
| $g$ | acceleration due to gravity | $m\,s^{-2}$ |
| $G$ | mass flow rate of fluid | $kg\,s^{-1}$ |
| $G_w$ | mass flow rate of gas under maximum (choked) flow rate conditions | $kg\,s^{-1}$ |
| $H$ | enthalpy of gas | $J\,kg^{-1}$ |
| $H_1$ | enthalpy of gas at section 1 | $J\,kg^{-1}$ |
| $H_w$ | enthalpy of gas at the section where flow rate is a maximum | $J\,kg^{-1}$ |
| $K$ | a constant | $J\,kg^{-1}$ |
| $l$ | pipe length | m |
| $M$ | molar mass of gas | $kg\,mol^{-1}$ |
| $P$ | pressure of gas | Pa |
| $P_1$ | pressure of gas at section 1 | Pa |
| $P_2$ | pressure of gas at section 2 | Pa |
| $P_w$ | pressure of gas at the section where flow rate is a maximum | Pa |
| $R$ | universal gas constant | $J\,mol^{-1}\,K^{-1}$ |
| $R_0$ | shear stress experienced by fluid at the pipe wall | $N\,m^{-2}$ |
| $Re$ | Reynolds number for fluid flow | |
| $Re_G$ | Reynolds number which would obtain for flow of the gas phase in two-phase flow if the gas were considered to flow alone through the pipe | |
| $Re_L$ | Reynolds number which would obtain for flow of the liquid phase in two-phase flow if the liquid were considered to flow alone through the pipe | |

| Symbol | Description | Unit |
|---|---|---|
| $T$ | temperature of the fluid | K |
| $T_1$ | temperature of the fluid at section 1 | K |
| $T_2$ | temperature of the fluid at section 2 | K |
| $T_w$ | temperature of the fluid at the section where flow rate is a maximum | K |
| $u$ | average velocity of fluid in a pipe | m s$^{-1}$ |
| $u_1$ | average velocity of fluid at section 1 | m s$^{-1}$ |
| $u_2$ | average velocity of fluid at section 2 | m s$^{-1}$ |
| $u_w$ | average velocity of fluid where flow rate is a maximum | m s$^{-1}$ |
| $U$ | internal energy of unit mass of fluid | J kg$^{-1}$ |
| $v$ | specific volume of the gas | m$^3$ kg$^{-1}$ |
| $v_1$ | specific volume of the gas at section 1 | m$^3$ kg$^{-1}$ |
| $v_2$ | specific volume of the gas at section 2 | m$^3$ kg$^{-1}$ |
| $v_w$ | specific volume of the gas at section where flow rate is a maximum | m$^3$ kg$^{-1}$ |
| $W_s$ | work done by unit mass of the fluid | J kg$^{-1}$ |
| $X$ | $(\Delta P_L/\Delta P_G)^{0.5}$ | |
| $\alpha$ | the kinetic energy correction factor | |
| $\gamma$ | ratio of heat capacities, $C_p/C_v$ | |
| $\delta q$ | infinitesimal quantity of heat added to the fluid | J |
| $\Delta h$ | difference in liquid levels in a manometer | m |
| $\Delta P$ | $P_2 - P_1$ | Pa |
| $\Delta P_G$ | pressure difference across a pipe which would obtain for gas flow in two-phase flow if the gas were considered to flow alone through the pipe | Pa |
| $\Delta P_L$ | pressure difference across a pipe which would obtain for liquid flow in two-phase flow if the liquid were considered to flow alone through the pipe | Pa |
| $\Delta P_{TPF}$ | pressure difference across a pipe for gas-liquid, two-phase flow | Pa |
| $\Delta z$ | difference in height between section 2 and section 1 | m |
| $\mu$ | viscosity of fluid | kg m$^{-1}$ s$^{-1}$ |
| $\rho$ | density of fluid | kg m$^{-3}$ |
| $\rho_2$ | density of fluid at section 2 | kg m$^{-3}$ |
| $\rho_w$ | density of fluid at section where flow rate is a maximum | kg m$^{-3}$ |
| $\phi$ | Stanton–Pannell friction factor | |
| $\phi_G$ | Stanton–Pannell friction factor which would obtain for gas flow in two-phase flow if the gas were considered to flow alone through the pipe | |
| $\phi_L$ | Stanton–Pannell friction factor which would obtain for liquid flow in two-phase flow if the liquid were considered to flow alone through the pipe | |
| $\Phi_G$ | factor which, when squared, relates $-\Delta P_{TPF}$ to $\Delta P_G$ | |

# 3 Velocity boundary layers

## 3.1 Worked Examples
### 3.1.1 Streamline flow over a flat plate

*Problem*

(i) Establish that water (density = 998 kg m$^{-3}$, viscosity = 1.002 × 10$^{-3}$ kg m$^{-1}$s$^{-1}$) in the boundary layer, which forms as the fluid flows over a flat plate, would be entirely in streamline motion if the plate were 48.0 cm long in the direction of flow and when the fluid beyond the boundary layer is flowing with a uniform velocity of 19.6 cm s$^{-1}$. ($Re_{x,\text{crit}} = 10^5$.) (ii) How thick would this boundary layer be 30 cm from the leading edge? (iii) What would be the water velocity in the middle of the boundary layer? (iv) What shear stress would be exerted on the plate by the flowing fluid 30 cm from the leading edge? (v) What would be the mean shear stress experienced over the whole plate under these conditions? (vi) If the plate were 2.6 m wide, what would be the total force experienced by the plate?

*Solution*

(i) Fluid in the boundary layer would be entirely in streamline motion if, at the trailing edge of the plate, $Re_x < Re_{x,\text{crit}}$.

At this point,
$$Re_x = \frac{\rho u_s x}{\mu}$$
$$= \frac{998 \times 19.6 \times 10^{-2} \times 48.0 \times 10^{-2}}{1.002 \times 10^{-3}}$$
$$= 9.37 \times 10^4$$

This is less than $10^5$, so fluid in the boundary layer would be in streamline motion over the entire flat plate.

(ii) The thickness of a streamline boundary layer may be obtained from (ref. 1 and cf. ref. 2)
$$\frac{\delta}{x} = 4.64 \, Re_x^{-0.5}$$

## Velocity boundary layers 51

Thus, at 30 cm from the leading edge, boundary layer thickness,

$$\delta = 4.64 \times 30 \times 10^{-2} \left( \frac{998 \times 19.6 \times 10^{-2} \times 30 \times 10^{-2}}{1.002 \times 10^{-3}} \right)^{-0.5} \text{ m}$$

$$= \underline{5.75 \text{ mm}}$$

(iii) To find fluid velocity at a point inside the boundary layer, an appropriate velocity profile equation is required. If it is assumed that $u_x [ = f(y)]$ can be expressed as a power series (ref. 3),

$$u_x = \frac{3u_s}{2} \cdot \frac{y}{\delta} - \frac{u_s}{2} \left( \frac{y}{\delta} \right)^3$$

In the middle of the boundary layer, $y/\delta = 0.5$

So,
$$u_x = \frac{3 \times 19.6 \times 10^{-2} \times 0.5}{2} - \frac{19.6 \times 10^{-2} \times 0.5^3}{2} \text{ m s}^{-1}$$

$$= \underline{13.5 \text{ cm s}^{-1}}$$

If an alternative assumption for the form of the velocity profile equation is used, we may write, (ref. 2)

$$u_x = 2u_s \cdot \frac{y}{\delta} - u_s \left( \frac{y}{\delta} \right)^2$$

Then, in the middle of the boundary layer,

$$u_x = (2 \times 19.6 \times 10^{-2} \times 0.5) - (19.6 \times 10^{-2} \times 0.5^2) \text{ m s}^{-1}$$

$$= \underline{14.7 \text{ cm s}^{-1}}$$

Thus the fluid velocity in the middle of the boundary layer is about 14 cm s$^{-1}$, the calculated value depending on the expression used to describe the velocity profile.

(iv) The shear stress experienced by the plate at a distance $x$ from the leading edge, (ref. 4 or cf. ref. 5)

$$R = 0.323 \, \rho u_s^2 \, Re_x^{-0.5}$$

At 30 cm from the leading edge,

$$R = 0.323 \times 998 (19.6 \times 10^{-2})^2$$
$$\times \left( \frac{998 \times 19.6 \times 10^{-2} \times 30 \times 10^{-2}}{1.002 \times 10^{-3}} \right)^{-0.5} \text{ kg m}^{-1} \text{s}^{-2}$$

$$= \underline{5.12 \times 10^{-2} \text{ N m}^{-2}}$$

## 52  Problems in fluid flow

(v) The mean shear stress experienced over the whole plate of length x, (ref. 6),

$$R_{ms} = 0.646 \, \rho u_s^2 \, Re_x^{-0.5}$$

Since the plate is 48.0 cm long,

$$R_{ms} = 0.646 \times 998 \, (19.6 \times 10^{-2})^2$$

$$\times \left( \frac{998 \times 19.6 \times 10^{-2} \times 48.0 \times 10^{-2}}{1.002 \times 10^{-3}} \right)^{-0.5} \text{kg m}^{-1}\text{s}^{-2}$$

$$= \underline{8.09 \times 10^{-2} \text{ N m}^{-2}}$$

(vi) The total force experienced by the plate

$$= 8.09 \times 10^{-2} \times 2.6 \times 48.0 \times 10^{-2} \text{ kg m s}^{-2}$$

$$= \underline{0.101 \text{ N}}$$

## 3.1.2  Turbulent flow over a flat plate

*Problem*

Consider a system in which air (pressure = 102.7 kPa, temperature = 18 °C, viscosity = $1.827 \times 10^{-5}$ kg m$^{-1}$ s$^{-1}$) is passing over a flat plate (1.65 m long in the direction of flow, 4.7 m wide) at a velocity of 18.4 m s$^{-1}$ at points outside the boundary layer. Find (i) the percentage of the surface over which there will exist a turbulent boundary layer if $Re_{x,\text{crit}} = 10^5$,
(ii) the thickness of the boundary layer 1.3 m away from the leading edge,
(iii) the velocity of air at the mid point of the boundary layer,
(iv) the thickness of the laminar sub-layer 1.3 m away from the leading edge,
(v) the velocity of air at the outer edge of the laminar sub-layer 1.3 m away from the leading edge,
(vi) the shear stress exerted on the plate 1.3 m away from the leading edge,
(vii) the mean shear stress experienced over the whole flat plate, and
(viii) the total drag force experienced by the plate.

Universal gas constant = 8.314 J mol$^{-1}$ K$^{-1}$
Molar mass of air = $28.8 \times 10^{-3}$ kg mol$^{-1}$

*Solution*

The boundary layer forms as shown in Figure 3.1.

If air is considered to behave as an ideal gas,

$$PV = n\mathcal{R}T$$

*Velocity boundary layers* 53

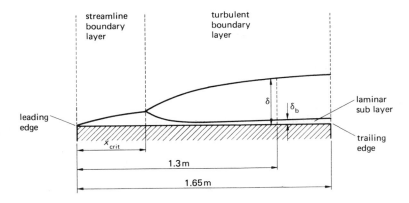

**Figure 3.1**

so that its density,
$$\frac{nM}{V} = \frac{PM}{\mathcal{R}T}$$

In this case then, gas density

$$= \frac{102.7 \times 10^3 \times 28.8 \times 10^{-3}}{8.314 \times 291} \text{ kg m}^{-3}$$

$$= 1.223 \text{ kg m}^{-3}$$

(i) The distance from the leading edge at which the boundary layer will become turbulent,

$$x_{crit} = \frac{Re_{x,crit}\,\mu}{\rho u_s}$$

$$= \frac{10^5 \times 1.827 \times 10^{-5}}{1.223 \times 18.4} = 0.0812 \text{ m}$$

Thus, the percentage of the surface over which a turbulent boundary layer will exist

$$= \frac{(1.65 - 0.0812)100}{1.65} \%$$

$$= 95.1 \%$$

Since this is a large fraction, for the purposes of the remaining calculations the boundary layer will be considered to be turbulent all the way from the leading edge. Thus:

(ii) the thickness of a turbulent boundary layer, (ref. 7 and 8),

$$\delta = 0.375\, Re_x^{-0.20}\, x$$

At 1.3 m away from the leading edge,

$$\delta = 0.375 \left(\frac{1.223 \times 18.4 \times 1.3}{1.827 \times 10^{-5}}\right)^{-0.20} \times 1.3 \text{ m}$$

$$= \underline{2.80 \text{ cm}}$$

(iii) The velocity of fluid inside a boundary layer is given by a velocity profile equation, which for a turbulent boundary layer is (ref. 9 and 10)

$$\frac{u_x}{u_s} = \left(\frac{y}{\delta}\right)^{1/7}$$

At the mid-points of the boundary layer, $y/\delta = 0.5$. So, the velocity of air at the mid-points,

$$u_x = 18.4 \times 0.5^{1/7} \text{ m s}^{-1}$$

$$= \underline{16.67 \text{ m s}^{-1}}$$

(iv) The thickness of the laminar sub-layer is given by (ref. 11):

$$\delta_b = 74.6 \, Re_x^{-0.9} \, x$$

So at 1.3 m away from the leading edge,

$$\delta_b = 74.6 \left(\frac{1.223 \times 18.4 \times 1.3}{1.827 \times 10^{-5}}\right)^{-0.9} \times 1.3 \text{ m}$$

$$= \underline{0.253 \text{ mm}}$$

(v) The velocity of air at the outer edge of the laminar sub-layer is given either by (ref. 11):

$$\frac{u_b}{u_s} = \left(\frac{\delta_b}{\delta}\right)^{1/7}$$

or

$$\frac{u_b}{u_s} = 2.13 \, Re_x^{-0.1}$$

Using the first of these equations, at 1.3 m away from the leading edge, velocity of air at the outer edge of the laminar sub-layer,

$$u_b = 18.4 \left(\frac{0.253 \times 10^{-3}}{2.80 \times 10^{-2}}\right)^{1/7} \text{ m s}^{-1}$$

$$= \underline{9.39 \text{ m s}^{-1}}$$

(vi) Now the shear stress experienced locally by the plate at a distance $x$ from the leading edge, (ref. 8 and 11),

$$R = 0.0286 \, \rho u_s^2 \, Re_x^{-0.2}$$

So, 1.3 m away from the leading edge,

$$R = 0.0286 \times 1.223 \times 18.4^2 \left(\frac{1.223 \times 18.4 \times 1.3}{1.827 \times 10^{-5}}\right)^{-0.2} \text{ kg m}^{-1}\text{s}^{-2}$$

$$= \underline{0.680 \text{ N m}^{-2}}$$

(vii) The mean shear stress experienced over the whole flat plate, (ref. 12),

$$R_{mt} = 0.0358 \, \rho u_s^2 \, Re_x^{-0.2}$$

So since the plate is 1.65 m long,

$$R_{mt} = 0.0358 \times 1.223 \times 18.4^2 \left(\frac{1.223 \times 18.4 \times 1.65}{1.827 \times 10^{-5}}\right)^{-0.2} \text{ kg m}^{-1}\text{s}^{-2}$$

$$= \underline{0.812 \text{ N m}^{-2}}$$

(viii) The total drag force experienced by the plate

$$= 0.812 \times 1.65 \times 4.7 \text{ kg m s}^{-2}$$

$$= \underline{6.30 \text{ N}}$$

## 3.1.3 Streamline and turbulent flow through a pipe, and equations of the universal velocity profile

*Problem*

Consider fully developed flow through a smooth pipe and equations of the universal velocity profile. If water (density = 998 kg m$^{-3}$, viscosity = 1.002 × 10$^{-3}$ kg m$^{-1}$ s$^{-1}$) flows through a pipe (I.D. = 3.2 cm) at a volumetric flow rate of 37.6 cm$^3$ s$^{-1}$, find (i) the value of pipe flow Reynolds number for fluid flow, (ii) the velocity gradient which will obtain in the water at the pipe wall, and (iii) the shear stress which will obtain there.

If the volumetric flow rate of water is now increased to 2.10 dm$^3$ s$^{-1}$, find (iv) the thickness of the laminar sub-layer, (v) the thickness of the buffer layer, (vi) the percentage of the pipe's cross-sectional area occupied by fully turbulent core, (vii) the velocity of fluid where laminar sub-layer and buffer layer meet, (viii) the velocity of fluid where buffer layer and turbulent core meet, (ix) the velocity of fluid along the pipe axis, and (x) the shear stress which will obtain at the pipe wall.

*Solution*

Consider the system with water flowing at 37.6 cm$^3$ s$^{-1}$.

Since volumetric flow rate, $Q = \dfrac{\pi d^2 u}{4}$

average fluid velocity, $u = \dfrac{4Q}{\pi d^2}$

$$= \dfrac{4 \times 37.6 \times 10^{-6}}{\pi (3.2 \times 10^{-2})^2} \text{ m s}^{-1} = 4.675 \text{ cm s}^{-1}$$

Now, (i) pipe-flow Reynolds number,

$$Re = \dfrac{998 \times 4.675 \times 10^{-2} \times 3.2 \times 10^{-2}}{1.002 \times 10^{-3}}$$

$$= \underline{1490}$$

So water will be in streamline motion in the pipe.

(ii) Where flow is laminar, (ref. 13 and 14),

$$u^+ = y^+$$

Now by definition, $\quad u^+ = \dfrac{u_x}{u^*}$

Also by definition, $\quad u^* = \left(\dfrac{R}{\rho}\right)^{0.5}$

so that $\quad = u\left(\dfrac{R}{\rho u^2}\right)^{0.5}$

$$= u\phi^{0.5}$$

So, $\quad u^+ = \dfrac{u_x}{u\phi^{0.5}}$

Also by definition, $\quad y^+ = \dfrac{\rho u^* y}{\mu}$

$$= \dfrac{y}{d} \cdot \dfrac{u^*}{u} \cdot \dfrac{\rho u d}{\mu}$$

$$= \dfrac{y}{d} \cdot \phi^{0.5} \cdot Re$$

So, $\quad \dfrac{u_x}{u\phi^{0.5}} = \dfrac{y}{d} \phi^{0.5} Re$

and $\quad u_x = \dfrac{y}{d} u\phi Re \qquad (3.1)$

## Velocity boundary layers

Expanding $\phi$ and $Re$ in terms of constituent factors,

$$u_x = \frac{y}{d} \cdot u \cdot \frac{R_o}{\rho u^2} \cdot \frac{\rho u d}{\mu}$$

$$= \frac{y R_o}{\mu}$$

But,
$$R_o = -\mu \frac{du_x}{dy}\bigg|_{y=0}$$

So,
$$u_x = -y \frac{du_x}{dy}\bigg|_{y=0} \tag{3.2}$$

Equating the right-hand sides of equations (3.1) and (3.2),

$$\frac{y}{d} u \phi Re = -y \frac{du_x}{dy}\bigg|_{y=0}$$

so,
$$\frac{du_x}{dy}\bigg|_{y=0} = -\frac{u \phi Re}{d}$$

Now for streamline flow, (ref. 15 and 16),

$$\phi = \frac{8}{Re}$$

So the velocity gradient at the pipe wall,

$$\frac{du_x}{dy}\bigg|_{y=0} = -\frac{8u}{d}$$

$$= -\frac{8 \times 4.675 \times 10^{-2}}{3.2 \times 10^{-2}} \text{ s}^{-1}$$

$$= \underline{-11.7 \text{ s}^{-1}}$$

(iii) The shear stress which will obtain at the pipe wall,

$$R_o = -\mu \frac{du_x}{dy}\bigg|_{y=0}$$

$$= 1.002 \times 10^{-3} \times 11.7 \text{ kg m}^{-1}\text{s}^{-2}$$

$$= \underline{1.17 \times 10^{-2} \text{ N m}^{-2}}$$

If the volumetric flow rate is now increased to 2.10 dm³ s⁻¹, average fluid velocity,

$$u = \frac{4 \times 2.10 \times 10^{-3}}{\pi (3.2 \times 10^{-2})^2} \text{ m s}^{-1}$$

$$= 2.611 \text{ m s}^{-1}$$

and pipe flow Reynolds number,
$$Re = \frac{998 \times 2.611 \times 3.2 \times 10^{-2}}{1.002 \times 10^{-3}}$$
$$= 8.32 \times 10^4$$

So now, streamline flow is confined to an annular laminar sub-layer, and since the pipeline is smooth, (ref. 17 and 18),
$$\phi = 0.0396 \, Re^{-0.25}$$
i.e.
$$\phi = 0.0396(8.32 \times 10^4)^{-0.25}$$
$$= 2.33 \times 10^{-3}$$

Now (iv) the thickness of the laminar sub-layer is given by, (ref. 19),
$$\frac{\delta_b}{d} = 5 \, Re^{-1} \, \phi^{-0.5}$$
i.e.
$$\delta_b = \frac{5 \times 3.2 \times 10^{-2}}{8.32 \times 10^4 (2.33 \times 10^{-3})^{0.5}} \, m,$$
$$= 39.8 \, \mu m$$

(v) The buffer layer extends inwards from the pipe wall until (ref. 14 and 19),
$$y^+ = \frac{y}{d} \phi^{0.5} \, Re \qquad \text{[see section (ii)]},$$
$$= 30$$
i.e.
$$y = \frac{30 \times 3.2 \times 10^{-2}}{(2.33 \times 10^{-3})^{0.5} \times 8.32 \times 10^4} \, m$$
$$= 0.239 \, mm$$

So the thickness of the buffer layer
$$= (0.239 \times 10^{-3}) - (39.8 \times 10^{-6}) \, m$$
$$= \underline{0.199 \, mm}$$

(vi) Now the pipe's cross-sectional area
$$= \frac{\pi (3.2 \times 10^{-2})^2}{4} \, m^2$$
$$= 8.042 \times 10^{-4} \, m^2$$

## Velocity boundary layers 59

The diameter of the turbulent core
$$= (3.2 \times 10^{-2}) - 2(0.239 \times 10^{-3}) \text{ m}$$
$$= 3.152 \text{ cm}$$

The cross-sectional area of the turbulent core
$$= \frac{\pi(3.152 \times 10^{-2})^2}{4} \text{ m}^2$$
$$= 7.803 \times 10^{-4} \text{ m}^2$$

So the percentage of the pipe's cross-sectional area occupied by the turbulent core
$$= \frac{7.803 \times 10^{-4} \times 100}{8.042 \times 10^{-4}} \%$$
$$= \underline{97.0\%}$$

(vii) Where the laminar sub-layer and buffer layer meet, (ref. 19 and cf. ref. 20),
$$u^+ = y^+ = 5$$

But, $\quad u^+ = \dfrac{u_x}{u\phi^{0.5}} \quad$ [see section (ii)]

So the velocity of fluid there,
$$u_x = 5 \times 2.611(2.33 \times 10^{-3})^{0.5} \text{ m s}^{-1}$$
$$= \underline{0.630 \text{ m s}^{-1}}$$

(viii) Where the buffer layer and turbulent core meet, (ref. 19 and 20),
$$y^+ = 30$$
But for the turbulent core, (ref. 14 and 19),
$$u^+ = 2.5 \ln y^+ + 5.5$$

i.e. $\quad \dfrac{u_x}{u\phi^{0.5}} = 2.5 \ln y^+ + 5.5. \quad$ [see section (ii)]

Thus, fluid velocity here,
$$u_x = 2.611(2.33 \times 10^{-3})^{0.5}(2.5 \ln 30 + 5.5) \text{ m s}^{-1}$$
$$= \underline{1.76 \text{ m s}^{-1}}$$

(ix) Universal velocity profile equations give erroneous values towards the pipe axis because of assumptions made during their derivation. But using the

Prandtl $\frac{1}{7}$th Power Law, it may be shown that (ref. 21 and 22):

$$u = 0.81\, u_s$$

So fluid velocity along the pipe axis,

$$u_s = \frac{2.611}{0.81}\ \text{m s}^{-1}$$

$$= \underline{3.22\ \text{m s}^{-1}}$$

(x) Finally, the shear stress, $R_o$, which will obtain at the pipe wall is given by the appropriate equation due to Blasius, (refs. 17 and 18),

i.e.       $\dfrac{R_o}{\rho u^2} = 0.0396\, Re^{-0.25} = \phi$

Thus,       $R_o = 2.33 \times 10^{-3} \times 998 \times 2.611^2\ \text{kg m}^{-1}\text{s}^{-2}$

$$= \underline{15.85\ \text{N m}^{-2}}$$

### 3.1.4 Streamline flow through a pipe when flow patterns are not fully developed

*Problem*

Consider a liquid which passes in streamline motion through the entire length of a pipe when the pipe is too short for fully developed flow to be established before liquid emerges from the discharge end. Where $\delta$ is the thickness to which the boundary layer has developed at the discharge end of the pipe, derive an expression for the total volumetric flow rate of liquid passing through the pipe.

*Solution*

Flow rate will be determined by velocity distribution in the discharge-end cross section because the boundary layer increases in thickness progressively all the way to the discharge end.

In the discharge-end cross section, volumetric flow rate through the boundary-layer annulus will depend upon velocities given by a velocity profile equation for streamline boundary layers (ref. 3),

e.g.       $u_x = \dfrac{3u_s}{2} \cdot \dfrac{y}{\delta} - \dfrac{u_s}{2}\left(\dfrac{y}{\delta}\right)^3$

Fluid will be moving in plug flow, at velocity $u_s$, in the central core. So total

volumetric flow rate,
$$Q_{total} = Q_{bl} + Q_{core} \qquad (3.3)$$
Consider fluid in the discharge-end cross section as shown in Figure 3.2.

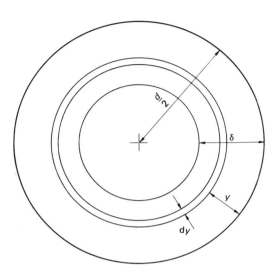

**Figure 3.2**

In equation (3.3), $\qquad Q_{core} = \pi\left(\dfrac{d}{2} - \delta\right) u_s$

Also, the area of a circular element of cross section

$$= \pi\left(\frac{d}{2} - y\right)^2 - \pi\left(\frac{d}{2} - y - dy\right)^2$$

$$= \pi(d - 2y)\,dy$$

So, $\qquad Q_{bl} = \displaystyle\int_0^\delta \pi(d - 2y) u_x \cdot dy$

$$= \pi \int_0^\delta (d - 2y)\left(\frac{3u_s}{2\delta} \cdot y - \frac{u_s}{2\delta^3} \cdot y^3\right) dy$$

$$= \pi \int_0^\delta \left(\frac{3u_s d}{2\delta} \cdot y - \frac{3u_s}{\delta} \cdot y^2 - \frac{u_s d}{2\delta^3} \cdot y^3 + \frac{u_s}{\delta^3} \cdot y^4\right) dy$$

$$= \pi \left[\frac{3u_s d}{4\delta} \cdot y^2 - \frac{u_s}{\delta} \cdot y^3 - \frac{u_s d}{8\delta^3} \cdot y^4 + \frac{u_s}{5\delta^3} \cdot y^5\right]_0^\delta$$

$$= \pi \left( \frac{3u_s d}{4} \cdot \delta - u_s \delta^2 - \frac{u_s d}{8} \cdot \delta + \frac{u_s}{5} \cdot \delta^2 \right)$$

$$= \frac{5\pi u_s d}{8} \cdot \delta - \frac{4\pi u_s}{5} \cdot \delta^2$$

Thus, the total volumetric flow rate,

$$Q_{total} = \frac{5\pi u_s d}{8} \cdot \delta - \frac{4\pi u_s}{5} \cdot \delta^2 + \pi u_s \left( \frac{d}{2} - \delta \right)^2$$

$$= \frac{\pi u_s d^2}{4} - \frac{3\pi u_s \delta d}{8} + \frac{\pi u_s \delta^2}{5}$$

## 3.2 Student Exercises

1 (i) To what distance from the leading edge could water in a boundary layer remain entirely in streamline motion as it flowed over a flat surface if velocity of fluid in the main stream beyond the boundary layer were 15.5 cm s$^{-1}$? ($Re_{x,\,crit} = 10^5$.)

(ii) How thick would the boundary layer become before turbulence is initiated in the boundary layer?

(iii) With what velocity would water be moving in the middle of streamline parts of the boundary layer?

(iv) What would be the local shear stress experienced by the surface 24 cm from the leading edge?

(v) What would be the mean shear stress experienced by the surface under the streamline part of the boundary layer?

Density of water = 998 kg m$^{-3}$
Viscosity of water = $1.002 \times 10^{-3}$ kg m$^{-1}$ s$^{-1}$

2 A thin flat plate (1.25 m wide, 2.77 m long) is being propelled through stagnant water at a velocity of 2.16 m s$^{-1}$.

Find: (i) the percentage of the surfaces over which boundary layers will be turbulent if $Re_{x,\,crit} = 10^5$,

(ii) the thickness to which boundary layers will have developed at the trailing edges of the plate,

(iii) the velocity, measured with respect to an external fixed point, of water at mid-points of the boundary layers,

(iv) the thickness to which laminar sub-layers will have developed at the trailing edges of the plate,

(v) the local shear stress exerted under each boundary layer half way along the plate,

(vi) the mean shear stress experienced by the plate,

(vii) the total force necessary to maintain plate velocity at 2.16 m s$^{-1}$.

*Velocity boundary layers* 63

Density of water = 998 kg m$^{-3}$
Viscosity of water = $1.002 \times 10^{-3}$ kg m$^{-1}$ s$^{-1}$

3  Consider fully developed flow through a pipeline and equations of the universal velocity profile. An oil (density = 923 kg m$^{-3}$, viscosity = $1.97 \times 10^{-2}$ kg m$^{-1}$ s$^{-1}$) flows through a pipeline (I.D. = 15.0 cm). Find:
(i) the maximum volumetric flow rate at which flow would remain streamline, ($Re_{crit} = 2100$),
(ii) the velocity gradient which will then obtain at the pipe wall, and
(iii) the shear stress which will then obtain there.
When pipe-flow Reynolds number = 2100, find also:
(iv) the velocity of fluid along the pipe axis, and
(v) the velocity of fluid 1 cm away from the pipe wall.

4  Consider fully developed flow through a smooth pipeline and equations of the universal velocity profile. If oil (density = 923 kg m$^{-3}$, viscosity = $1.97 \times 10^{-2}$ kg m$^{-1}$ s$^{-1}$) flows through a pipeline (I.D. = 15.0 cm) at 62.8 dm$^3$ s$^{-1}$,
(i) establish that the fluid stream will then be turbulent. Then find:
(ii) the value of the Stanton–Pannell friction factor ($\phi$) for the system,
(iii) the thickness of the laminar sub-layer,
(iv) the thickness of the buffer layer,
(v) the velocity of fluid where turbulent core and buffer layer meet,
(vi) the velocity gradient and shear stress where turbulent core and buffer layer meet, and
(vii) the shear stress at the pipe wall.

## 3.3 References

References are abbreviated as follows:

C & R (1) = J. M. Coulson and J. F. Richardson, *Chemical Engineering*, Vol. 1, 3rd edition, Pergamon, 1977

D, G & S = J. F. Douglas, J. M. Gasiorek and J. A. Swaffield, *Fluid Mechanics*, Pitman, 1980

1. C & R (1), page 312,
2. D, G & S, page 296
3. C & R (1), page 311
4. C & R (1), page 313
5. D, G & S, page 298
6. C & R (1), page 314
7. C & R (1), page 316
8. D, G & S, page 300
9. C & R (1), page 315
10. D, G & S, page 299
11. C & R (1), page 317
12. C & R (1), page 318
13. C & R (1), page 325
14. D, G & S, page 264
15. C & R (1), page 53
16. D, G & S, page 256
17. C & R (1), page 42
18. D, G & S, page 255
19. C & R (1), page 326
20. D, G & S, page 263
21. C & R (1), page 56
22. D, G & S, page 266

## 3.4 Notation

| Symbol | Description | Unit |
|---|---|---|
| $d$ | pipe diameter | m |
| $M$ | molar mass of a gas | kg mol$^{-1}$ |
| $n$ | number of moles of a gas | |
| $P$ | pressure of gas | Pa |
| $Q$ | volumetric flow rate of fluid | m$^3$ s$^{-1}$ |
| $Q_{bl}$ | volumetric flow rate of fluid through the boundary layer in a pipe | m$^3$ s$^{-1}$ |
| $Q_{core}$ | volumetric flow rate of fluid in the central core of fluid beyond the boundary layer when fully developed flow has not become established in a pipe | m$^3$ s$^{-1}$ |
| $Q_{total}$ | total volumetric flow rate of fluid through a pipe | m$^3$ s$^{-1}$ |
| $\mathscr{R}$ | universal gas constant | J mol$^{-1}$ K$^{-1}$ |
| $R$ | shear stress experienced by a flat plate under a stream of fluid, and at a point distance $x$ from the leading edge | N m$^{-2}$ |
| $R_{ms}$ | mean value of shear stress experienced by a flat plate under a streamline boundary layer | N m$^{-2}$ |
| $R_{mt}$ | mean value of shear stress experienced by a flat plate under a turbulent boundary layer | N m$^{-2}$ |
| $R_o$ | shear stress experienced at a pipe wall with fluid flow being fully developed in the pipe | N m$^{-2}$ |
| $Re$ | Reynolds number for fluid flow in a pipe | |
| $Re_x$ | Reynolds number for fluid flow over a flate plate, written with $x$ as length dimension | |
| $Re_{x,crit}$ | the critical value of $Re_x$ at which fluid flow first becomes turbulent | |
| $u$ | average velocity of fluid in a pipe | m s$^{-1}$ |
| $u_b$ | fluid velocity at the surface of the laminar sub-layer remote from the bounding wall | m s$^{-1}$ |
| $u_s$ | velocity of fluid in the main stream beyond the boundary layer | m s$^{-1}$ |
| $u_x$ | velocity of fluid in the $x$ direction and in the boundary layer at a point distance $y$ from the surface over which fluid is flowing | m s$^{-1}$ |
| $u^*$ | shearing stress velocity $[=(R/\rho)^{0.5}]$ | m s$^{-1}$ |
| $u^+$ | $u_x/u^*$ | |
| $V$ | volume occupied by a gas | m$^3$ |
| $x$ | distance from the leading edge of a surface | m |
| $x_{crit}$ | distance from the leading edge at which fluid flow in the boundary layer first becomes turbulent | m |
| $y$ | distance from the surface over which fluid is flowing | m |
| $y^+$ | $(\rho u^* y)/\mu$ | |
| $\delta$ | thickness of the boundary layer | m |
| $\delta_b$ | thickness of the laminar sub layer | m |
| $\mu$ | fluid viscosity | kg m$^{-1}$ s$^{-1}$ |
| $\rho$ | fluid density | kg m$^{-3}$ |
| $\phi$ | Stanton–Pannell friction factor $(=R_o/\rho u^2)$ | |

# 4 Flow measurement

## 4.1 Worked Examples

### 4.1.1 Use of a Pitot tube to measure flow rate in fluid streams in fully developed flow

*Problem*

Calculate maximum fluid velocity ($\dot{u}_{max}$), Reynolds number ($\rho \dot{u}_{max} d/\mu$), and the volumetric and mass flow rates (in $m^3 s^{-1}$ and $kg s^{-1}$ respectively) of fluids in the following situations, while passing in fully developed flow through pipe lines of inside diameter = 54.0 cm, and when axial fluid velocity is measured by using a Pitot tube:
(i) water, when the vertical difference in mercury levels in a manometer coupled to the Pitot tube = 28.6 cm, and leads between the Pitot tube and the manometer are filled with water, and
(ii) natural gas (pressure = 689 kPa, temperature = 21 °C), when the gas flows isothermally, the vertical difference in water levels in a manometer coupled to the Pitot tube = 17.4 cm, and natural gas fills the leads between the Pitot tube and the manometer.

Density of water = 998 $kg\,m^{-3}$
Density of mercury = $1.354 \times 10^4$ $kg\,m^{-3}$
Average molar mass of the natural gas = $2.83 \times 10^{-2}$ $kg\,mol^{-1}$
Viscosity of water = $1.001 \times 10^{-3}$ $kg\,m^{-1}\,s^{-1}$
Viscosity of the natural gas = $1.182 \times 10^{-5}$ $kg\,m^{-1}\,s^{-1}$
Universal gas constant = 8.314 $J\,K^{-1}\,mol^{-1}$
Acceleration due to gravity = 9.81 $m\,s^{-2}$

*Solution*

(i) The axial velocity will be the maximum velocity of fluid in the pipeline, so since water is incompressible, (ref. 1, 2 and 3),
$$\dot{u}_{max} = [2v(P_2 - P_1)]^{0.5}$$

In this equation, specific volume of the water,
$$v = 1/998 = 1.002 \times 10^{-3} \text{ m}^3 \text{ kg}^{-1}$$
$$P_2 - P_1 = \Delta h(\rho_{Hg} - \rho_w)g$$
$$= 28.6 \times 10^{-2} \times 9.81(1.354 \times 10^4 - 998) \text{ Pa} = 35.189 \text{ kPa}$$

So $\dot{u}_{max} = (2 \times 1.002 \times 10^{-3} \times 35.189 \times 10^3)^{0.5} \text{ m s}^{-1}$

$= \underline{8.40 \text{ m s}^{-1}}$

Reynolds number, $\rho \dot{u}_{max} d/\mu = \dfrac{998 \times 8.40 \times 54.0 \times 10^{-2}}{1.001 \times 10^{-3}}$

$= \underline{4.52 \times 10^6}$

Turbulence, therefore, is well developed, so that average fluid velocity, (ref. 4, 5 and 6),
$$u = 0.81 \dot{u}_{max}$$
$$= 0.81 \times 8.40 = 6.804 \text{ m s}^{-1}$$

So, volumetric flow rate of water

$$= \dfrac{\pi(54.0 \times 10^{-2})^2 \times 6.804}{4} \text{ m}^3 \text{ s}^{-1}$$

$$= \underline{1.558 \text{ m}^3 \text{ s}^{-1}}$$

and mass flow rate $= 1.558 \times 998 \text{ kg s}^{-1}$

$= \underline{1555 \text{ kg s}^{-1}}$

(ii) The axial (maximum) fluid velocity for a compressible fluid such as natural gas, (ref. 1),
$$\dot{u}_{max} = [2P_1 v_1 \ln (P_2/P_1)]^{0.5}$$
Now if the gas behaves ideally and flows isothermally,
$$Pv = \dfrac{RT}{M} = \text{constant} = P_1 v_1$$

Thus, $P_1 v_1 = \dfrac{8.314 \times 294}{2.83 \times 10^{-2}} = 8.637 \times 10^4 \text{ J kg}^{-1}$

Since gas pressure, $P_1 = 689 \times 10^3 \text{ Pa}$

specific volume of gas in the system,

$$v_1 = \dfrac{8.637 \times 10^4}{689 \times 10^3} = 0.1254 \text{ m}^3 \text{ kg}^{-1}$$

*Flow measurement* 67

and gas density, $\rho_g = 1/0.1254 = 7.977 \text{ kg m}^{-3}$

Now, $\quad P_2 - P_1 = \Delta h(\rho_w - \rho_g)g$

$\quad\quad\quad = 17.4 \times 10^{-2} \times 9.81(998 - 7.977) = 1690 \text{ kg m}^{-1}\text{s}^{-2}$

So, impact pressure at the tip of the Pitot tube,

$$P_2 = P_1 + 1690 \text{ kg m}^{-1}\text{s}^{-2}$$
$$= (689 \times 10^3) + 1690 = 6.907 \times 10^5 \text{ Pa}$$

Hence,

$$\dot{u}_{max} = \{2 \times 8.637 \times 10^4 [\ln(6.907 \times 10^5)/(6.89 \times 10^5)]\}^{0.5} \text{ m s}^{-1}$$
$$= \underline{20.63 \text{ m s}^{-1}}$$

Reynolds number,

$$\rho \dot{u}_{max} d/\mu = \frac{7.977 \times 20.63 \times 54.0 \times 10^{-2}}{1.182 \times 10^{-5}}$$
$$= \underline{7.518 \times 10^6}$$

So again, turbulent flow is well developed, so that average fluid velocity in the pipe, (ref. 4, 5 and 6),

$$u = 0.81 \times 20.63 \text{ m s}^{-1}$$
$$= 16.71 \text{ m s}^{-1}$$

Volumetric flow rate of natural gas is, therefore,

$$= \frac{\pi(54.0 \times 10^{-2})^2 \times 16.71}{4} \text{ m}^3 \text{ s}^{-1}$$
$$= \underline{3.827 \text{ m}^3 \text{ s}^{-1}}$$

and mass flow rate $\quad = 3.827 \times 7.977 \text{ kg s}^{-1}$
$$= \underline{30.53 \text{ kg s}^{-1}}$$

## 4.1.2 Use of a Pitot tube to determine flow rate in a gas stream in which flow patterns are not fully developed

*Problem*

A measure of the flow rate of chlorine gas (pressure = 148 kPa, temperature = 16 °C) is required in a pipeline (inside diameter = 35.0 cm) which is too short for fully developed flow to have been established in it. Data obtained by

## 68  Problems in fluid flow

traversing the pipeline with a Pitot tube connected to a water-filled manometer are given in Table 4.1. Use the data to evaluate graphically volumetric and mass flow rates of chlorine in the pipeline.

**Table 4.1**

| Radial distance from pipe wall (cm) | 1 | 2.5 | 5 | 10 | 15 | 17.5 |
|---|---|---|---|---|---|---|
| Difference in liquid levels in manometer (cm) | 0.20 | 0.36 | 0.54 | 0.81 | 0.98 | 1.00 |

Molar mass of chlorine gas = $7.09 \times 10^{-2}$ kg mol$^{-1}$
Universal gas constant = $8.314$ J K$^{-1}$ mol$^{-1}$
Acceleration due to gravity = $9.81$ m s$^{-2}$
Density of water = $999$ kg m$^{-3}$

*Solution*

The volumetric flow rate of fluid through a pipeline of circular cross section, (cf. ref. 4, 7 and 8),

$$Q = \int_0^{r_w} \dot{u}\, dA$$

$$= 2\pi \int_0^{r_w} \dot{u} r\, dr \qquad (4.1)$$

So to find $Q$ when fluid velocity, $\dot{u}$, is not a simple function of $r$, the integral in equation (4.1) may be evaluated graphically by measuring the area under a curve obtained by plotting $\dot{u} r$ versus $r$ between $r = 0$ and $r = r_w$.

Now for a compressible fluid like chlorine gas, (ref. 1)

$$\dot{u} = [2P_1 v_1 \ln(P_2/P_1)]^{0.5} \qquad (4.2)$$

and $$P_2 - P_1 = \Delta h (\rho_w - \rho_{Cl_2}) g \qquad (4.3)$$

(ref. 9, 10 and 11)

In these equations,

$$\rho_{Cl_2} = \frac{P_1 M}{RT}$$

and $$P_1 v_1 = \frac{RT}{M}$$

Substituting data, $P_1v_1 = \dfrac{8.314 \times 289}{7.09 \times 10^{-2}} = 3.389 \times 10^4 \text{ J kg}^{-1}$

$$\rho_{Cl_2} = \dfrac{148 \times 10^3 \times 7.09 \times 10^{-2}}{8.314 \times 289} \text{ kg m}^{-3}$$

$$= 4.367 \text{ kg m}^{-3}$$

So using equation (4.3),

$$P_2 = [148 \times 10^3 + (999 - 4.367) \times 9.81 \times \Delta h] \text{ Pa}$$

$$= (1.48 \times 10^5 + 9.757 \times 10^3 \, \Delta h) \text{ Pa} \qquad (4.4)$$

Equations (4.2) and (4.4) may now be used to calculate $\dot{u}$ values at the various positions in the pipe for which data have been given. These data and values of the product $\dot{u}r$ are given in Table 4.2.

**Table 4.2**

| Radial distance from pipe wall (cm) | $r$ (m) | $\dot{u}$ (m s$^{-1}$) | $\dot{u}r$ (m$^2$ s$^{-1}$) |
|---|---|---|---|
| 0.0 | 0.175 | 0.0 | 0.000 |
| 1.0 | 0.165 | 2.989 | 0.493 |
| 2.5 | 0.150 | 4.011 | 0.602 |
| 5.0 | 0.125 | 4.912 | 0.614 |
| 10.0 | 0.075 | 6.015 | 0.451 |
| 15.0 | 0.025 | 6.616 | 0.165 |
| 17.5 | 0.0 | 6.684 | 0.000 |

Now by drawing a graph of $\dot{u}r$ versus $r$, and measuring the area under the curve between $r = 0$ and $r = 0.175$ m, it may be shown that

$$\int_0^{0.175} \dot{u}r \, dr = 0.0726 \text{ m}^3 \text{ s}^{-1}$$

So, volumetric flow rate of the chlorine gas,

$$Q = 2\pi \times 0.0726 \text{ m}^3 \text{ s}^{-1}$$

$$= \underline{0.456 \text{ m}^3 \text{ s}^{-1}}$$

Since gas density, $\rho_{Cl_2} = 4.367 \text{ kg m}^{-3}$

the mass flow rate of chlorine gas

$$= 0.456 \times 4.367 \text{ kg s}^{-1}$$

$$= \underline{1.99 \text{ kg s}^{-1}}$$

## 4.1.3 Use of an orifice plate and manometer to measure fluid flow rate in horizontal and vertical pipelines

*Problem*

(i) An orifice plate (orifice diameter = 12.4 cm) is set in a horizontal pipeline (inside diameter = 15.0 cm) which will carry an oil of density = 877 kg m$^{-3}$. If a water manometer, with leads filled with the oil, is coupled across the orifice plate, what will be the maximum rate (in kg s$^{-1}$) at which oil can be metered through the pipeline if the difference in water levels in the manometer must not exceed 75 cm?

(ii) Suppose now that the orifice plate were set in a vertical part of the same pipeline system with oil flowing upwards through it at the maximum flow rate calculated above. If the tapping points for the manometer leads were 1.0 and 0.5 times the pipe diameter upstream and downstream of the orifice plate respectively, calculate the new difference in pressures which would occur between the tapping points, and also what the difference in water levels would now be in the manometer.

Coefficient of discharge = 0.61
Density of water = 999 kg m$^{-3}$
Acceleration due to gravity = 9.81 m s$^{-2}$

*Solution*

(i) In this system, the orifice occupies a large fraction of the total cross-sectional area of the pipeline, so we must use (ref. 12, 13 and 14):

$$G = \frac{C_D A_o}{v} \left( \frac{2[g(z_1 - z_2) + v(P_1 - P_2)]}{1 - (A_o/A_1)^2} \right)^{0.5} \quad (4.5)$$

However, the orifice plate is set firstly in a horizontal part of the pipeline, so $(z_1 - z_2) = 0$.

Also, 
$$\frac{A_o}{A_1} = \left(\frac{d_o}{d_1}\right)^2$$

so equation (4.5) reduces to:

$$G = \frac{C_D A_o}{v} \left( \frac{2v(P_1 - P_2)}{1 - (d_o/d_1)^4} \right)^{0.5} \quad (4.6)$$

Now, 
$$P_1 - P_2 = \Delta h (\rho_w - \rho_{oil}) g$$

(refs. 9, 10 and 11)

The maximum pressure difference which may be generated, therefore,

$$= 75 \times 10^{-2} (999 - 877) \times 9.81 = 897.6 \text{ Pa}$$

Also, specific volume of the oil,
$$v = 1/\rho_{oil}$$
$$= 1/877 = 1.140 \times 10^{-3} \text{ m}^3 \text{ kg}^{-1}$$

So using equation (4.6), the maximum mass flow rate at which oil may be metered,

$$G = \frac{0.61 \times \pi \times (12.4 \times 10^{-2})^2}{4 \times 1.140 \times 10^{-3}} \times \left( \frac{2 \times 1.140 \times 10^{-3} \times 897.6}{1 - \left(\frac{12.4 \times 10^{-2}}{15.0 \times 10^{-2}}\right)^4} \right)^{0.5} \text{ kg s}^{-1}$$

$$= \underline{12.66 \text{ kg s}^{-1}}$$

(ii) When the orifice plate is installed in the vertical pipe with oil flowing upwards through it,

$$z_1 - z_2 = -[(1.0 + 0.5) \times 15.0 \times 10^{-2}] \text{ m}$$
$$= -0.225 \text{ m}$$

Rearranging equation (4.5), we find:

$$v(P_1 - P_2) = \frac{1}{2}\left(\frac{Gv}{C_D A_o}\right)^2 \left[1 - \left(\frac{d_o}{d_1}\right)^4\right] - g(z_1 - z_2)$$

So by substituting data, and since $v = 1/\rho_{oil}$, the difference in pressures which would now occur between tapping points,

$$P_1 - P_2 = \frac{877}{2}\left(\frac{12.66 \times 1.140 \times 10^{-3} \times 4}{0.61 \times \pi (12.4 \times 10^{-2})^2}\right)^2 \left[1 - \left(\frac{12.4}{15.0}\right)^4\right]$$
$$+ 9.81 \times 877 \times 0.225 \text{ Pa}$$

$$= \underline{2833 \text{ Pa}}$$

But as shown in Figure 4.1, the manometer leads leave the pipeline at different levels.

The pressure at point X due to $P_1$ and fluids in the left-hand leg of the manometer $= P_1 + (\Delta h + h_{oil})\rho_{oil} g + h_w \rho_w g$.

The pressure at point X due to $P_2$ and fluids in the right-hand leg of the manometer $= P_2 + [h_{oil} + (z_2 - z_1)]\rho_{oil} g + (\Delta h + h_w)\rho_w g$.

At steady state, these two pressures must be equal,

i.e., 
$$P_1 + \Delta h \rho_{oil} g + h_{oil} \rho_{oil} g + h_w \rho_w g$$
$$= P_2 + h_{oil} \rho_{oil} g + (z_2 - z_1)\rho_{oil} g + \Delta h \rho_w g + h_w \rho_w g$$

Rearranging,
$$P_1 - P_2 = \Delta h(\rho_w - \rho_{oil})g + (z_2 - z_1)\rho_{oil} g$$

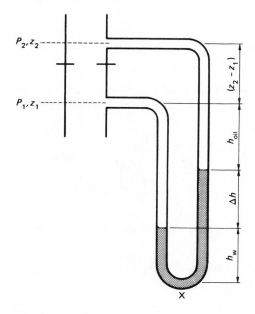

**Figure 4.1**

Now substituting data,

$$2833 = \Delta h(999 - 877)9.81 + 0.225 \times 877 \times 9.81$$

from which, the difference in water levels in the manometer,

$$\underline{\Delta h = 75.0 \text{ cm}}$$

(*Note*: The difference in liquid levels in a manometer coupled across a constriction in a pipeline does not depend upon inclination of the pipeline into which the constriction is inserted.)

### 4.1.4 Determination of orifice size required to measure a flow rate of fluid with an orifice meter, and comparison of pressure drops produced by orifice and venturi meters

*Problem*

(i) Determine the diameter of a sharp-edged orifice needed in an orifice plate inserted into a horizontal pipeline (inside diameter = 6.5 cm) in order to obtain a 20 cm difference in mercury levels in a manometer coupled across the plate when 24% aqueous sodium hydroxide solution flows through the pipeline at 16.5 dm³ s⁻¹, and the manometer leads are filled with the sodium hydroxide solution.

Density of mercury = $1.356 \times 10^4$ kg m$^{-3}$
Density of 24% aqueous NaOH = 1266 kg m$^{-3}$
Discharge coefficient = 0.61
Acceleration due to gravity = 9.81 m s$^{-2}$

(ii) Determine the ratio of pressure drop readings to be expected by using firstly a horizontal orifice meter, then a horizontal venturi meter, to measure the same flow rates of liquid in a pipe when the hole in the orifice plate and the throat of the venturi meter have the same diameter.

Discharge coefficient for the orifice meter = 0.61
Discharge coefficient for the venturi meter = 0.98

*Solution*

(i) The mass flow rate of an incompressible fluid as it flows through an orifice meter is given (from ref. 12, 13 and 14) by:

$$G = \frac{C_D A_o}{v} \left( \frac{2[g(z_1 - z_2) + v(P_1 - P_2)]}{1 - (A_o/A_1)^2} \right)^{0.5} \tag{4.7}$$

If the pipe in which the orifice plate is inserted is horizontal, $(z_1 - z_2) = 0$, and equation (4.7) may be rearranged to give:

$$A_o = \frac{Gv}{C_D} \left( \frac{1}{2v(P_1 - P_2) + (Gv/C_D A_1)^2} \right)^{0.5} \tag{4.8}$$

In this case, the mass flow rate of NaOH solution,

$$G = Q\rho_{NaOH}$$
$$= 16.5 \times 10^{-3} \times 1266 = 20.89 \text{ kg s}^{-1}$$

Specific volume of NaOH solution,

$$v = 1/\rho_{NaOH}$$
$$= 1/1266 = 7.899 \times 10^{-4} \text{ m}^3 \text{ kg}^{-1}$$

$$A_1 = \frac{\pi (6.5 \times 10^{-2})^2}{4} = 3.318 \times 10^{-3} \text{ m}^2$$

$$P_1 - P_2 = \Delta h (\rho_{Hg} - \rho_{NaOH}) g$$
$$= 20 \times 10^{-2} (1.356 \times 10^4 - 1266) 9.81 \text{ Pa}$$
$$= 2.412 \times 10^4 \text{ Pa}$$

Now, substituting in equation (4.8), the area of the orifice required

$$A_o = \frac{20.89 \times 7.899 \times 10^{-4}}{0.61}$$

$$\times \left(1 \bigg/ \left[(2 \times 7.899 \times 10^{-4} \times 2.412 \times 10^4) + \left(\frac{20.89 \times 7.899 \times 10^{-4}}{0.61 \times 3.318 \times 10^{-3}}\right)^2\right]\right)^{0.5} \text{m}^2$$

$$= 2.645 \times 10^{-3} \text{ m}^2$$

So the diameter of orifice needed

$$= \left(\frac{4 \times 2.645 \times 10^{-3}}{\pi}\right)^{0.5} \text{m}$$

$$= \underline{5.80 \text{ cm}}$$

(ii) Let subscripts o and v to $C_D$ and $\Delta P$ signify values referring to orifice meter and venturi meter systems respectively. Then, where the two meter systems are arranged horizontally, and for the same liquid flow rate, from equation (4.7), we have for the venturi meter:

$$\left(\frac{Gv}{C_{D,v} A_o}\right)^2 = \frac{2v\Delta P_v}{(A_o/A_1)^2 - 1} \qquad (4.9)$$

and for the orifice meter:

$$\left(\frac{Gv}{C_{D,o} A_o}\right)^2 = \frac{2v\Delta P_o}{(A_o/A_1)^2 - 1} \qquad (4.10)$$

Now by dividing equation (4.10) by equation (4.9) and simplifying, the pressure drop ratio required,

$$\frac{\Delta P_o}{\Delta P_v} = \left(\frac{C_{D,v}}{C_{D,o}}\right)^2$$

$$= \left(\frac{0.98}{0.61}\right)^2$$

$$= \underline{2.58}$$

### 4.1.5 Use of a rotameter to measure fluid flow rates

*Problem*

A rotameter consists of a ceramic float (density = 3778 kg m$^{-3}$, volume = 0.92 cm$^3$) riding in a vertical, tapered tube (length = 25.0 cm, inside

diameter at the top = 12.5 mm, inside diameter at the bottom = 9.5 mm). The greatest diameter of the float = 9.5 mm. With argon (pressure = 115 kPa, temperature = 16 °C) passing through the meter, calculate:
(i) the pressure drop in the gas stream which must occur over the float, and
(ii) the mass and volumetric flow rates of argon when the float takes up a position with its maximum diameter 7.6 cm from the top of the tapered tube.

Molar mass of argon = $3.995 \times 10^{-2}$ kg mol$^{-1}$
Acceleration due to gravity = 9.81 m s$^{-2}$
Universal gas constant = 8.314 J K$^{-1}$ mol$^{-1}$
Discharge coefficient = 0.94

*Solution*

(i) When a float takes up an equilibrium position in a rotameter tube, the net downward force acting on the float due to gravity is balanced by the upward force due to fluid drag. The net downward force

= (mass of float − mass of fluid displaced) × acceleration due to gravity

= $V_f(\rho_f - \rho)g$

Thus, the pressure drop sustained by fluid passing the float,

$$-\Delta P = \frac{V_f(\rho_f - \rho)g}{A_f} \qquad (4.11)$$

where $A_f$ is the maximum cross-sectional area of the float in a horizontal plane. If the gas behaves ideally,

$$Pv = \frac{RT}{M}$$

so fluid density, $\qquad \rho = 1/v = \dfrac{PM}{RT}$

$$= \frac{115 \times 10^3 \times 3.995 \times 10^{-2}}{8.314 \times 289} = 1.912 \text{ kg m}^{-3}$$

Also, $\qquad A_f = \dfrac{\pi(9.5 \times 10^{-3})^2}{4}$ m$^2$

$= 7.088 \times 10^{-5}$ m$^2$

Now substituting data into equation (4.11), the pressure drop over the float,

$$-\Delta P = \frac{(0.92 \times 10^{-6})(3778 - 1.912) \times 9.81}{7.088 \times 10^{-5}} \text{ Pa}$$

$= \underline{480.8 \text{ Pa}}$

# 76 Problems in fluid flow

(ii) The pressure drop across the float is very small compared to the total gas pressure in the system, so we may consider the fluid to behave incompressibly. Also, if the vertical height of the float is neglected, mass flow rate of argon through the system, (from ref. 12, 13 and 14),

$$G = C_D A_o \left( \frac{2\rho(-\Delta P)}{1 - (A_o/A_1)^2} \right)^{0.5} \qquad (4.12)$$

in which $A_o$ is the cross-sectional area of the annular orifice between float and tube, and $A_1$ is the cross-sectional area of the tube at the level of the float.

Now consider the tapered tube shown in section in Figure 4.2.

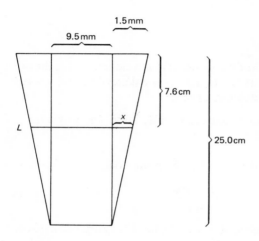

**Figure 4.2**

From similar triangles, distance $x$ at level $L$ is given by:

$$\frac{x}{25.0 - 7.6} = \frac{0.15}{25.0}$$

from which $\qquad x = 0.1044$ cm

Thus the tube diameter at level $L = 0.95 + 2 \times 0.1044$ cm,

$$= 1.1588 \text{ cm}$$

Hence, $\qquad A_1 = \dfrac{\pi (1.1588 \times 10^{-2})^2}{4} = 1.0546 \times 10^{-4}$ m$^2$

and $\qquad A_o = A_1 - A_f$

$$= (1.0546 \times 10^{-4}) - (7.088 \times 10^{-5}) \text{ m}^2$$

$$= 3.458 \times 10^{-5} \text{ m}^2$$

Now substituting data into equation (4.12), mass flow rate of argon,

$$G = 0.94 \times 3.458 \times 10^{-5} \left( \frac{2 \times 1.912 \times 480.8}{1 - \left(\frac{3.458 \times 10^{-5}}{1.0546 \times 10^{-4}}\right)^2} \right)^{0.5} \text{ kg s}^{-1}$$

$$= \underline{1.475 \times 10^{-3} \text{ kg s}^{-1}}$$

Volumetric flow rate of argon

$$= \frac{1.475 \times 10^{-3}}{1.912} \text{ m}^3 \text{ s}^{-1}$$

$$= \underline{0.771 \text{ dm}^3 \text{ s}^{-1}}$$

### 4.1.6 Mass of a float required to measure fluid flow rate with a rotameter

*Problem*

A steel float (greatest diameter = 14.2 mm) is to ride inside a tapered rotameter tube (length = 280 mm, inside diameter at the top = 18.8 mm, inside diameter at the bottom = 14.2 mm). The float is required to occupy a position 70 mm from the top of the tube when water is passing through at 4.0 dm³ min⁻¹. Determine what the mass of the float must be if conditions are such that the discharge coefficient = 0.97.

Density of water = 999 kg m⁻³
Density of the steel = 8020 kg m⁻³
Acceleration due to gravity = 9.81 m s⁻²

*Solution*

If the height of the float is considered negligibly small, mass flow rate of an incompressible fluid such as water, (from ref. 12, 13 and 14)

$$G = \frac{C_D A_o}{v} \left( \frac{2v(-\Delta P)}{1 - (A_o/A_1)^2} \right)^{0.5} \quad (4.13)$$

in which $\quad -\Delta P = \dfrac{V_f (\rho_f - \rho) g}{A_f} \quad$ (see equation 4.11) $\quad (4.14)$

In this system, $\quad G = \dfrac{4.0 \times 10^{-3} \times 999}{60} = 0.0666 \text{ kg s}^{-1}$

and $\quad v = 1/\rho$

$\quad\quad\quad\quad = 1/999 = 1.001 \times 10^{-3} \text{ m}^3 \text{ kg}^{-1}$

The greatest cross-sectional area of the float,

$$A_f = \frac{\pi(14.2 \times 10^{-3})^2}{4} = 1.584 \times 10^{-4} \text{ m}^2$$

Now consider a vertical section of the rotameter tube as shown in Figure 4.3, in which the float must settle at level L.

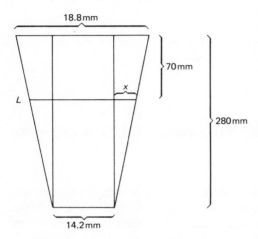

**Figure 4.3**

Considering similar triangles,

$$\frac{x}{280-70} = \frac{0.5(18.8-14.2)}{280}$$

from which, $\quad x = 1.725$ mm

So the tube diameter at level L $= 14.2 + (2 \times 1.725) = 17.65$ mm

and $\quad A_1 = \dfrac{\pi(17.65 \times 10^{-3})^2}{4} = 2.447 \times 10^{-4}$ m²

Thus, the area of the annular orifice,

$$A_o = A_1 - A_f$$
$$= 2.447 \times 10^{-4} - 1.584 \times 10^{-4} \text{ m}^2$$
$$= 8.630 \times 10^{-5} \text{ m}^2$$

Now from equation (4.14) and equation (4.13) rearranged,

$$\frac{V_f(\rho_f - \rho)g}{A_f} = \frac{1}{2v}\left(\frac{Gv}{C_D A_o}\right)^2 \left[1 - \left(\frac{A_o}{A_1}\right)^2\right]$$

So, substituting data, the volume of float required,

$$V_f = \frac{1.584 \times 10^{-4}}{2 \times 1.001 \times 10^{-3} \times 9.81(8020 - 999)} \times \left(\frac{0.0666 \times 1.001 \times 10^{-3}}{0.97 \times 8.630 \times 10^{-5}}\right)^2$$

$$\times \left[1 - \left(\frac{0.8630}{2.447}\right)^2\right] m^3$$

$$= 6.380 \times 10^{-7} \, m^3$$

Since the density of float material $= 8020 \, kg \, m^{-3}$

the mass of float required $= 6.380 \times 10^{-7} \times 8020 \, kg$

$$= \underline{5.12 \, g}$$


## 4.2 Student Exercises

1  What will be the difference in heights of liquid levels in a manometer coupled to a Pitot tube which is located along the axis of a pipe
(i) if water (density $= 998 \, kg \, m^{-3}$) flows through the pipe (inside diameter $= 26.5 \, cm$) at a volumetric flow rate $= 160.7 \, dm^3 \, s^{-1}$, when the manometer fluid is mercury (density $= 1.355 \times 10^4 \, kg \, m^{-3}$) and leads to the manometer are filled with water, and
(ii) if air (pressure $= 232 \, kPa$, temperature $= 19 \,°C$) flows through the pipe (inside diameter $= 46.5 \, cm$) at $16.8 \, kg \, s^{-1}$, when the manometer fluid is oil (density $= 905 \, kg \, m^{-3}$) and air fills the leads between the Pitot tube and the manometer.

Average molar mass of air $= 2.88 \times 10^{-2} \, kg \, mol^{-1}$
Viscosity of air $= 1.78 \times 10^{-5} \, kg \, m^{-1} \, s^{-1}$
Viscosity of water $= 1.002 \times 10^{-3} \, kg \, m^{-1} \, s^{-1}$
Universal gas constant $= 8.314 \, J \, K^{-1} \, mol^{-1}$
Acceleration due to gravity $= 9.81 \, m \, s^{-2}$


2  Flow rate of oil (density $= 882 \, kg \, m^{-3}$) must be measured in a pipeline (inside diameter $= 60 \, cm$) close to a valve where a disturbed flow pattern obtains in the fluid. Data, obtained by traversing the pipeline with a Pitot tube connected to a manometer filled with an aqueous sodium chloride solution (density $= 1203 \, kg \, m^{-3}$), are given in Table 4.3. If the manometer leads are filled with the oil, use the data to determine volumetric flow rate of oil in the pipeline.

Acceleration due to gravity $= 9.81 \, m \, s^{-2}$

## Table 4.3

| Radial distance from pipe wall (cm) | 1 | 2 | 5 | 10 | 18 | 24 | 30 |
|---|---|---|---|---|---|---|---|
| Average difference in liquid levels in manometer (cm) | 8.2 | 14.6 | 16.1 | 23.8 | 26.2 | 25.4 | 25.9 |

3   The flow rate of acetone (density = 789.9 kg m$^{-3}$) in a pipeline (inside diameter = 5.4 cm) is to be measured by inserting an orifice plate into a horizontal portion of the pipeline, and coupling an inclined, mercury-filled manometer across tapping points upstream and downstream of the orifice plate. If the manometer leads are filled with acetone, the vertical difference in mercury levels in the manometer cannot exceed 8.0 cm, and the maximum volumetric flow rate of the acetone will be 3.60 dm$^3$ s$^{-1}$, find the diameter of the orifice which should be cut into the orifice plate.

Coefficient of discharge = 0.61
Density of mercury = 1.356 × 10$^4$ kg m$^{-3}$
Acceleration due to gravity = 9.81 m s$^{-2}$

4   A horizontal pipeline (inside diameter = 18.0 cm) is fitted with an orifice plate, with an orifice diameter = 6.5 cm, and a mercury filled manometer across the pressure tappings. If methane gas (upstream pressure = 464 kPa, temperature = 25 °C) passes through the pipeline at 1.265 kg s$^{-1}$, what will be the difference in mercury levels observed in the manometer if the manometer leads are filled with methane?

Molar mass of methane = 1.603 × 10$^{-2}$ kg mol$^{-1}$
Viscosity of methane gas = 1.17 × 10$^{-5}$ kg m$^{-1}$ s$^{-1}$
Density of mercury = 1.3534 × 10$^4$ kg m$^{-3}$
Coefficient of discharge = 0.61
Universal gas constant = 8.314 J K$^{-1}$ mol$^{-1}$
Acceleration due to gravity = 9.81 m s$^{-2}$

5   A rotameter consists of a float (density = 4120 kg m$^{-3}$, volume = 1.24 cm$^3$) riding in a vertical, tapered tube (length = 25.0 cm, inside diameter at the top = 1.25 cm, inside diameter at the bottom = 0.95 cm). The float has its greatest diameter at the top, and this diameter = 0.95 cm. With carbon dioxide (pressure = 138 kPa, temperature = 4 °C) passing through the meter, and when the system has settled to a steady state, calculate:
(i) the pressure drop in the gas stream which will occur over the float, and
(ii) the distance from the top of the tube at which the top of the float will ride when the carbon dioxide is flowing at 1.08 dm$^3$ s$^{-1}$.

Molar mass of carbon dioxide = $4.4 \times 10^{-2}$ kg mol$^{-1}$
Discharge coefficient = 0.93
Universal gas constant = 8.314 J K$^{-1}$ mol$^{-1}$
Acceleration due to gravity = 9.81 m s$^{-2}$

6  A hollowed aluminium float (greatest diameter = 14.2 mm, overall volume = 8.20 cm$^3$) is to ride inside a tapered rotameter tube (length = 280 mm, inside diameter at the top = 18.8 mm, inside diameter at the bottom = 14.2 mm). The float is required to occupy a position 120 mm from the bottom of the tube when argon (pressure = 113 kPa, temperature = 16 °C) is passing through at 1.35 dm$^3$ s$^{-1}$. Determine what percentage of the overall volume of the float must be hollowed cavity if the mass of argon it accommodates and displaces is neglected.

Molar mass of argon = $3.995 \times 10^{-2}$ kg mol$^{-1}$
Density of aluminium = 2713 kg m$^{-3}$
Discharge coefficient = 0.91
Universal gas constant = 8.314 J K$^{-1}$ mol$^{-1}$
Acceleration due to gravity = 9.81 m s$^{-2}$

## 4.3  References

References are abbreviated as follows:

C & R (1) = J. M. Coulson and J. F. Richardson, Chemical Engineering, Vol. 1, 3rd edition, Pergamon, 1977

McC, S & H = W. L. McCabe, J. C. Smith and P. Harriott, Unit Operations of Chemical Engineering, 4th edition, McGraw-Hill, 1985

D, G & S = J. F. Douglas, J. M. Gasiorek and J. A. Swaffield, Fluid Mechanics, Pitman, 1980

1. C & R (1), page 102
2. McC, S & H, page 200
3. D, G & S, page 168
4. C & R (1), page 56
5. McC, S & H, page 86
6. D, G & S, page 266
7. McC, S & H, page 84
8. D, G & S, page 265
9. C & R (1), page 99
10. McC, S & H, page 29
11. D, G & S, page 44
12. C & R (1), page 105
13. McC, S & H, page 194
14. D, G & S, page 171

## 4.4  Notation

| Symbol | Description | Unit |
|---|---|---|
| $A$ | cross-sectional area of a pipe | m$^2$ |
| $A_1$ | cross-sectional area of a pipe at section 1 | m$^2$ |

## Problems in fluid flow

| Symbol | Description | Unit |
|---|---|---|
| $A_f$ | maximum cross-sectional area of a float | $m^2$ |
| $A_o$ | cross-sectional area of an orifice | $m^2$ |
| $C_D$ | discharge coefficient | |
| $C_{D,o}$ | discharge coefficient for an orifice meter | |
| $C_{D,v}$ | discharge coefficient for a venturi meter | |
| $d$ | inside diameter of a pipe | m |
| $d_1$ | inside diameter of a pipe at section 1 | m |
| $d_o$ | diameter of an orifice | m |
| $g$ | acceleration due to gravity | $m\,s^{-2}$ |
| $G$ | mass flow rate | $kg\,s^{-1}$ |
| $h_{oil}$ | a distance defined in Figure 4.1 | m |
| $h_w$ | a distance defined in Figure 4.1 | m |
| $M$ | molar mass of a gas | $kg\,mol^{-1}$ |
| $P$ | pressure of a fluid | Pa |
| $P_1$ | fluid pressure at section 1 | Pa |
| $P_2$ | fluid pressure at section 2 | Pa |
| $Q$ | volumetric flow rate | $m^3\,s^{-1}$ |
| $r$ | distance from the axis of a pipe | m |
| $r_w$ | inside radius of a pipe | m |
| $R$ | universal gas constant | $J\,K^{-1}\,mol^{-1}$ |
| $T$ | gas temperature | K |
| $u$ | average fluid velocity in a pipe | $m\,s^{-1}$ |
| $\dot{u}$ | local fluid velocity | $m\,s^{-1}$ |
| $\dot{u}_{max}$ | local maximum (axial) fluid velocity | $m\,s^{-1}$ |
| $v$ | specific volume of fluid | $m^3\,kg^{-1}$ |
| $v_1$ | specific volume of fluid at section 1 | $m^3\,kg^{-1}$ |
| $V_f$ | volume of a float | $m^3$ |
| $x$ | a distance defined in Figures 4.2 and 4.3 | m |
| $z_1$ | distance above a datum level at section 1 | m |
| $z_2$ | distance above a datum level at section 2 | m |
| $\Delta h$ | difference in height of liquid levels in a manometer | m |
| $\Delta P$ | $P_2 - P_1$ | Pa |
| $\Delta P_o$ | $P_2 - P_1$ for an orifice meter | Pa |
| $\Delta P_v$ | $P_2 - P_1$ for a venturi meter | Pa |
| $\mu$ | fluid viscosity | $kg\,m^{-1}\,s^{-1}$ |
| $\rho$ | fluid density | $kg\,m^{-3}$ |
| $\rho_{Cl_2}$ | density of chlorine gas | $kg\,m^{-3}$ |
| $\rho_f$ | density of float material | $kg\,m^{-3}$ |
| $\rho_g$ | density of gas | $kg\,m^{-3}$ |
| $\rho_{Hg}$ | density of mercury | $kg\,m^{-3}$ |
| $\rho_{NaOH}$ | density of a sodium hydroxide solution | $kg\,m^{-3}$ |
| $\rho_{oil}$ | density of oil | $kg\,m^{-3}$ |
| $\rho_w$ | density of water | $kg\,m^{-3}$ |

# 5 Flow and flow measurement in open channels

## 5.1 Worked Examples

### 5.1.1 Use of the Manning and Chézy formulae to analyse steady, uniform flow

*Problem*

Use the Manning formula to calculate average velocity of water if it flows in a straight, open, concrete channel (Manning resistance coefficient, $n = 0.016 \text{ m}^{-0.333} \text{ s}$), 0.65 m wide and with a constant slope of 1 unit vertical to 6950 units horizontal when the water depth is constant at 32.6 cm. What will be the volumetric flow rate of the water? If the Chézy formula were to be used to describe flow in this system, deduce the value which would obtain for the Chézy coefficient ($C$). What would be the shear stress experienced by the walls of the channel under these circumstances? If viscosity of the water = $1.308 \times 10^{-3} \text{ kg m}^{-1} \text{s}^{-1}$, what would be the velocity gradient in the fluid at the channel wall?

Density of water = 999.7 kg m$^{-3}$
Acceleration due to gravity = 9.81 m s$^{-2}$

*Solution*

The Manning formula (ref. 1, 2 and 3) is:

$$u = (1/n) \, m^{\frac{2}{3}} s^{\frac{1}{2}}$$

Now the cross-sectional area of the fluid stream in the channel

$$A = 0.65 \times 32.6 \times 10^{-2} \text{ m}^2$$
$$= 0.2119 \text{ m}^2$$

The length of the wetted perimeter

$$P = 0.65 + (2 \times 32.6 \times 10^{-2}) \text{ m}$$
$$= 1.302 \text{ m}$$

So, hydraulic mean depth for the flow, (ref. 2 and 4),

$$m = \frac{0.2119}{1.302} = 0.1627 \text{ m}$$

Channel slope, $\quad s = 1/6950 = 1.439 \times 10^{-4}$ radians

So using the Manning formula, average fluid velocity,

$$u = \frac{0.1627^{\frac{2}{3}} \times (1.439 \times 10^{-4})^{\frac{1}{2}}}{0.016} \text{ m s}^{-1}$$

$$= \underline{0.2234 \text{ m s}^{-1}}$$

Volumetric flow rate of the water

$$= A \cdot u$$
$$= 0.2119 \times 0.2234 \text{ m}^3 \text{ s}^{-1}$$
$$= \underline{47.34 \text{ dm}^3 \text{ s}^{-1}}$$

Now the Chézy formula (ref. 1 and 3) is:

$$u = C m^{\frac{1}{2}} s^{\frac{1}{2}}$$

so

$$C = \frac{0.2234}{0.1627^{0.5} \times (1.439 \times 10^{-4})^{0.5}} \text{ m}^{0.5} \text{ s}^{-1}$$

$$= \underline{46.17 \text{ m}^{0.5} \text{ s}^{-1}}$$

From a force balance on fluid in steady, uniform flow, it may be shown that shear stress experienced by the channel walls (ref. 5),

$$R_m = m\rho g \sin s$$

Since the channel bed slope is very small

$$\sin s \approx s$$

so $\quad R_m = 0.1627 \times 999.7 \times 9.81 \times 1.439 \times 10^{-4} \text{ kg m}^{-1} \text{ s}^{-2}$

$$= \underline{0.2296 \text{ N m}^{-2}}$$

Finally, $\quad R_m = -\mu \left(\frac{du}{dy}\right)_{y=0} \quad$ (ref. 6 and 7)

so velocity gradient in the fluid at the channel wall,

$$\left(\frac{du}{dy}\right)_{y=0} = -\frac{0.2296}{1.308 \times 10^{-3}} \text{ s}^{-1}$$

$$= \underline{-175.5 \text{ s}^{-1}}$$

## 5.1.2 Stream depth in a channel of trapezoidal cross section

*Problem*

Use the Manning formula to calculate the depth at which water will flow in steady, uniform motion and at a rate of 0.885 m³ s⁻¹ in an iron channel of trapezoidal cross section which slopes downwards one unit vertically over a horizontal distance of 960 units. The channel will have a base width of 1.36 m and sides which slope outwards at 55° to the horizontal. The Manning resistance coefficient, $n$, which will obtain for the material of construction $= 0.014 \text{ m}^{-0.333}$ s.

*Solution*

The Manning formula (ref. 1, 2 and 3) is:

$$u = (1/n) m^{\frac{2}{3}} s^{\frac{1}{2}}$$

so volumetric flow rate,

$$Q = u \cdot A$$

$$= \frac{A}{n} m^{\frac{2}{3}} s^{\frac{1}{2}}$$

Now

$$m = A/P$$

so

$$Q = \frac{A^{\frac{5}{3}} s^{\frac{1}{2}}}{n P^{\frac{2}{3}}} \tag{5.1}$$

The channel in cross section will appear as in Figure 5.1.

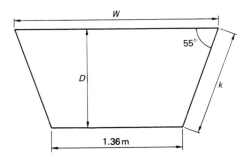

**Figure 5.1**

Now,

$$k = \frac{D}{\sin 55°}$$

and

$$W = 1.36 + \left(2 \times \frac{D}{\tan 55°}\right) \text{ m}$$

## 86 Problems in fluid flow

So cross sectional area,

$$A = D \left( \frac{1.36 + 1.36 + \dfrac{2D}{\tan 55°}}{2} \right) \text{ m}^2$$

$$= 1.36D + 0.7002D^2 \text{ m}^2$$

$$P = 1.36 + \frac{2D}{\sin 55°} \text{ m}$$

$$= 1.36 + 2.442D \text{ m}$$

Bed slope,

$$s = 1/960 = 1.042 \times 10^{-3} \text{ radians}$$

Now substituting data into equation (5.1)

$$0.885 = \frac{(1.042 \times 10^{-3})^{0.5}(1.36D + 0.7002D^2)^{\frac{5}{3}}}{0.014(1.36 + 2.442D)^{\frac{2}{3}}}$$

This equation may be solved graphically, or by trial and error. Hence, depth of the fluid stream, $\underline{D = 48.13 \text{ cm.}}$

### 5.1.3 Optimum base angle for a V-shaped channel. Slope of a channel

*Problem*

Find the base angle of a V-shaped channel which will make volumetric flow rate of liquid a maximum for a stream of given cross-sectional area.

At what slope should such a channel be set to conduct water at 25 dm³ s⁻¹ when the stream has a depth at the centre of 12 cm? The Manning resistance coefficient may be taken $= 0.011 \text{ m}^{-0.333} \text{ s}$.

*Solution*

Consider a channel with proportions as shown in Figure 5.2.

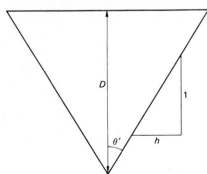

**Figure 5.2**

Using the Manning formula (ref. 1, 2 and 3), volumetric flow rate,

$$Q = (A/n) m^{2/3} s^{1/2}$$

$$= \frac{A^{5/3} s^{1/2}}{n P^{2/3}}$$

Using the Chézy formula (ref. 1 and 3), volumetric flow rate,

$$Q = AC m^{1/2} s^{1/2}$$

$$= \frac{C A^{3/2} s^{1/2}}{P^{1/2}}$$

So whichever formula is used, for a given cross-sectional area and channel slope, volumetric flow rate will be a maximum when $P$ is a minimum.

Now $\qquad P = 2(h^2 D^2 + D^2)^{0.5}$

and $\qquad A = h D^2$

so $\qquad P = 2\left(\dfrac{h^2 A}{h} + \dfrac{A}{h}\right)^{0.5}$

from which $\qquad P^2 = 4hA + \dfrac{4A}{h}$

Now if $P$ is a minimum, $P^2$ will be a minimum also. So for a given $A$,

$$\frac{d(P^2)}{dh} = 4A - \frac{4A}{h^2}$$

$P$ will be a minimum if $d(P^2)/dh = 0$

i.e. if $\qquad 4A = \dfrac{4A}{h^2}$

i.e. when $\qquad h = 1$

Now referring to Figure 5.2,

$$\tan \theta' = h$$

So for optimum base angle,

$$\tan \theta' = 1$$

i.e. $\qquad \theta' = 45°$

and optimum base angle, $\underline{2\theta' = 90°}$

Now $\qquad Q = \dfrac{A}{n}\left(\dfrac{A}{P}\right)^{2/3} s^{1/2}$ $\qquad$ (ref. 1, 2, 3 and 4)

from which $\qquad s = \dfrac{Q^2 n^2 P^{4/3}}{A^{10/3}}$ $\qquad$ (5.2)

## 88 Problems in fluid flow

In this case, with a base angle to the channel of 90° and a fluid depth of 12 cm,

$$A = (12 \times 10^{-2})^2 = 0.0144 \text{ m}^2$$

and
$$P = 2[(12 \times 10^{-2})^2 + (12 \times 10^{-2})^2]^{0.5} \text{ m}$$
$$= 0.3394 \text{ m}$$

So substituting data into equation (5.2),

$$\text{channel slope}, s = \frac{(25 \times 10^{-3})^2 \times 0.011^2 \times 0.3394^{1.333}}{0.0144^{3.333}} \text{ radians}$$

$$= 0.0246 \text{ radians}$$
$$= 1.41 \text{ degrees}$$
$$= \underline{1 \text{ unit vertical to 40.65 units horizontal}}$$

### 5.1.4 Stream depth for maximum velocity and for maximum volumetric flow rate in a pipe

*Problem*

Using the Chézy formula, find the stream depth (as a fraction of pipe diameter) that is required to obtain (i) maximum velocity, then (ii) maximum volumetric flow rate in a pipe of circular cross section. What will be (iii) the maximum velocity, and (iv) the velocity at maximum volumetric flow rate, which will obtain when storm water flows in a closed conduit of circular cross section, if the conduit is 2 m in diameter and is inclined downwards at 0.35 degrees? Take the coefficient in the Chézy formula = $78.6 \text{ m}^{0.5} \text{ s}^{-1}$.

*Solution*

According to the Chézy formula, (ref. 1 and 3),

$$u = C m^{\frac{1}{2}} s^{\frac{1}{2}}$$

$$= C \left(\frac{A}{P}\right)^{\frac{1}{2}} s^{\frac{1}{2}} \quad (5.3)$$

so that
$$Q = C \left(\frac{A^3}{P}\right)^{\frac{1}{2}} s^{\frac{1}{2}} \quad (5.4)$$

Thus velocity will be maximum when $A/P$ is a maximum, whereas volumetric flow rate will be maximum when $A^3/P$ is a maximum.

Now consider the conduit as shown in section in Figure 5.3.

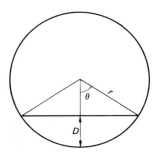

**Figure 5.3**

Firstly,
$$A = \theta r^2 - 2 \cdot \tfrac{1}{2} r \sin \theta \cdot r \cos \theta$$
$$= \theta r^2 - \tfrac{1}{2} r^2 \sin 2\theta \tag{5.5}$$

then
$$P = 2\theta r \tag{5.6}$$

From these,
$$\frac{dA}{d\theta} = r^2 - r^2 \cos 2\theta \tag{5.7}$$

and
$$\frac{dP}{d\theta} = 2r \tag{5.8}$$

(i) Now velocity will be maximum when $A/P$ is a maximum,

i.e. when
$$\frac{d(A/P)}{d\theta} = 0$$

Now
$$\frac{d(A/P)}{d\theta} = \frac{1}{P} \cdot \frac{dA}{d\theta} - \frac{A}{P^2} \cdot \frac{dP}{d\theta}$$

So using equations (5.5) to (5.8) to substitute, velocity will be maximum when:

$$\frac{1}{2\theta r}(r^2 - r^2 \cos 2\theta) = \frac{(\theta r^2 - \tfrac{1}{2} r^2 \sin 2\theta)}{4\theta^2 r^2} \cdot 2r$$

Simplifying, $\quad 2\theta = \tan 2\theta$

from which $\quad \theta = 2.247$ radians

Then, $\quad D = r - r \cos 2.247$

$\quad = r(1 + 0.626)$

$\quad = \underline{0.813 \times \text{pipe diameter}}$

So velocity will be maximum when stream depth = $0.813 \times$ pipe diameter.

## 90 Problems in fluid flow

(ii) Volumetric flow rate will be maximum when $A^3/P$ is a maximum,

i.e. when
$$\frac{\mathrm{d}(A^3/P)}{\mathrm{d}\theta} = 0$$

Now
$$\frac{\mathrm{d}(A^3/P)}{\mathrm{d}\theta} = \frac{3A^2}{P} \cdot \frac{\mathrm{d}A}{\mathrm{d}\theta} - \frac{A^3}{P^2} \cdot \frac{\mathrm{d}P}{\mathrm{d}\theta}$$

Using equations (5.5) to (5.8) to substitute, flow rate will be maximum when:

$$\frac{3(\theta r^2 - \tfrac{1}{2}r^2 \sin 2\theta)^2 (r^2 - r^2 \cos 2\theta)}{2\theta r} = \frac{(\theta r^2 - \tfrac{1}{2}r^2 \sin 2\theta)^3}{4\theta^2 r^2} \cdot 2r$$

Simplifying, $\quad 4\theta - 6\theta \cos 2\theta + \sin 2\theta = 0$

from which $\quad\quad\quad\quad\quad\quad\quad\quad \theta = 2.689$ radians

Then, $\quad\quad\quad\quad\quad\quad\quad\quad\quad D = r - r \cos 2.689$

$$= r(1 + 0.899)$$

$$= \underline{0.950 \times \text{pipe diameter}}$$

So volumetric flow rate will be maximum when stream depth = $0.950 \times$ pipe diameter

(iii) $\quad\quad\quad\quad\quad\quad\quad\quad u = C\left(\frac{A}{P}\right)^{0.5} s^{0.5}$

For maximum velocity, $\theta = 2.247$ radians

Using equation (5.5), $\quad A = 2.247 \times 1 - 0.5 \sin(2 \times 2.247) \text{ m}^2$

$$= 2.735 \text{ m}^2$$

Using equation (5.6), $\quad P = 2 \times 2.247 \times 1 \text{ m}$

$$= 4.494 \text{ m}$$

Now slope of the channel, $\quad s = 0.35$ degrees

$$= 0.006\ 109 \text{ radians}$$

So the maximum velocity of fluid which could obtain

$$= \frac{78.6 \times 2.735^{0.5} \times 0.006\ 109^{0.5}}{4.494^{0.5}} \text{ m s}^{-1}$$

$$= \underline{4.793 \text{ m s}^{-1}}$$

(iv) When maximum flow rate obtains,

$$\theta = 2.689 \text{ radians}$$

So using equation (5.5),

$$A = 2.689 \times 1 - 0.5 \sin(2 \times 2.689) \text{ m}^2$$
$$= 3.082 \text{ m}^2$$

Using equation (5.6),  $P = 2 \times 2.689 \times 1$ m
$$= 5.378 \text{ m}$$

So substituting data into the Chézy formula, fluid velocity at maximum volumetric flow rate

$$= \frac{78.6 \times 3.082^{0.5} \times 0.006\,109^{0.5}}{5.378^{0.5}} \text{ m s}^{-1}$$

$$= \underline{4.65 \text{ m s}^{-1}}$$

## 5.1.5 Flow measurement with sharp-crested weirs

*Problem*

(i) Calculate the velocity with which liquid will issue from the bottom of a notch cut into a sharp-crested weir if fluid pours freely over the weir, fluid velocity upstream is small enough to be neglected, and fluid at the notch is 28.0 cm deep.

(ii) If upstream fluid velocity is small enough to be neglected, what will be the depth of water at a rectangular notch, 46.0 cm wide, cut into a sharp-edged weir if fluid pours freely over the weir at 0.355 m$^3$ s$^{-1}$, (a) if no allowance is made for end contractions at the weir, and (b) if the Francis formula is used to allow for the notch being narrower than the main channel and situated symmetrically across the end of it?

(iii) What would be the base angle necessary for a V-shaped notch in a sharp-crested weir if liquid flowing at 22 dm$^3$ s$^{-1}$ is required to have a maximum depth of 18.5 cm in the notch, and upstream velocity may be neglected?

Coefficient of discharge in cases above $= 0.62$
Acceleration due to gravity $= 9.81$ m s$^{-2}$

*Solution*

(i) The velocity with which fluid at a depth, $h'$, below the top surface will issue from a notch in a sharp-crested weir is given by (ref. 8):

$$u_2 = (2gh')^{0.5}$$

So in this case,  $u_2 = (2 \times 9.81 \times 28.0 \times 10^{-2})^{0.5}$ m s$^{-1}$
$$= \underline{2.344 \text{ m s}^{-1}}$$

(ii) (a) If no allowance is made for end contractions at the weir, volumetric flow rate through the notch (ref. 8),

$$Q = \tfrac{2}{3} C_D B (2g)^{0.5} H^{1.5}$$

Thus,
$$H^{1.5} = \frac{3Q}{2C_D B (2g)^{0.5}}$$

from which fluid depth over the weir in this case,

$$H = \left( \frac{3 \times 0.355}{2 \times 0.62 \times 46.0 \times 10^{-2} (2 \times 9.81)^{0.5}} \right)^{\tfrac{2}{3}} \text{m}$$

$$= \underline{0.562 \text{ m}}$$

(b) If S.I. units are used, the Francis formula states that (ref. 9):

$$Q = 1.84 (B - 0.1 \, n' \, H) H^{1.5} \text{ m}^3 \text{ s}^{-1}$$

In this case, the number of end contractions, $n' = 2$.
So substituting data,

$$0.355 = 1.84 (46.0 \times 10^{-2} - 0.1 \times 2 \times H) H^{1.5}$$

This equation may be solved graphically, or by trial and error.

Hence, $\qquad H = 0.720 \text{ m or } 1.944 \text{ m}$

The nearest of these to the depth given unequivocally in part (a) is 0.720 m, so the fluid depth which may be expected over the weir, allowing for end contractions = $\underline{0.720 \text{ m}}$

(iii) The volumetric flow rate through a V-shaped notch (ref. 10).

$$Q = \tfrac{8}{15} C_D \tan \theta' (2g)^{0.5} H^{2.5}$$

Hence,
$$\tan \theta' = \frac{15 \, Q}{8 C_D (2g)^{0.5} H^{2.5}}$$

Substituting data,

$$\tan \theta' = \frac{15 \times 22 \times 10^{-3}}{8 \times 0.62 (2 \times 9.81)^{0.5} \times (18.5 \times 10^{-2})^{2.5}}$$

$$= 1.020$$

Hence, $\qquad \theta' = 45.57°$

so that the base angle of the notch would have to

$$= 2 \times 45.57 \text{ degrees}$$

$$= \underline{91.14°}$$

Flow and flow measurement in open channels 93

## 5.1.6 Equation for specific energy and analysis of streams in tranquil and shooting flow

*Problem*

What will be the specific energy of a stream of fluid, flowing in a channel of rectangular section and width 1.65 m, if volumetric flow rate of the fluid $= 0.675 \text{ m}^3 \text{s}^{-1}$ and stream depth $= 19.5$ cm? What will be the alternative depth at which fluid could flow with the same volumetric flow rate and specific energy? Is the first stream in tranquil or shooting flow? What is the maximum volumetric flow rate which could obtain at the specific energy of this system, and how (qualitatively) could the initial system be adjusted to obtain this maximum flow rate? In which direction and at what velocity with respect to the channel walls would small surface waves move in the initial system? What is the Froude number representative of the initial system? What percentage of the specific energy is due to kinetic energy in the initial system, and when fluid is flowing at the alternative depth?

Acceleration due to gravity $= 9.81 \text{ m s}^{-2}$

*Solution*

The specific energy of a fluid stream, when defined as energy per unit weight of fluid, is given by (ref. 11 and 12):

$$E = D + \frac{u^2}{2g}$$

Now for the stream in question, average fluid velocity,

$$u = \frac{Q}{BD}$$

$$= \frac{0.675}{1.65 \times 19.5 \times 10^{-2}} \text{ m s}^{-1}$$

$$= 2.098 \text{ m s}^{-1}$$

and flow rate per unit width,

$$q = \frac{Q}{B}$$

$$= \frac{0.675}{1.65} \text{ m}^2 \text{ s}^{-1}$$

$$= 0.4091 \text{ m}^2 \text{ s}^{-1}$$

So specific energy of the fluid,

$$E = 19.5 \times 10^{-2} + \frac{2.098^2}{2 \times 9.81} \text{ m}$$

$$= \underline{0.4193 \text{ m}}$$

Now
$$E = D + \frac{u^2}{2g}$$

$$= D + \frac{q^2}{2gD^2}$$

At the specific energy obtaining in the system,

$$0.4193 = D + \frac{0.4091^2}{2 \times 9.81 \times D^2}$$

Graphically, it may be shown that this equation has positive roots

$$D = 0.195 \text{ m and } 0.349 \text{ m}$$

So the alternative depth of flow at the specific energy and volumetric flow rate of this system = $\underline{0.349 \text{ m.}}$

Since the first stream is shallowest, it must be in <u>shooting flow</u>.

Now if $E$ is fixed, critical depth, $D_c$, will be given by (ref. 13, 14 and 15):

$$D_c = \tfrac{2}{3}E$$

Thus,
$$D_c = \frac{2 \times 0.4193}{3} \text{ m}$$

$$= 0.2795 \text{ m}$$

But
$$q_{max} = (gD_c^3)^{0.5} \qquad \text{(ref. 13 and 15)}$$

So maximum volumetric flow rate,

$$Q_{max} = B(gD_c^3)^{0.5}$$

$$= 1.65(9.81 \times 0.2795^3)^{0.5} \text{ m}^3\text{ s}^{-1}$$

$$= \underline{0.764 \text{ m}^3\text{ s}^{-1}}$$

Qualitatively, to obtain this maximum flow rate, the slope of the bed would have to be <u>decreased</u>, as dictated by the Manning formula. (The Chézy formula should not be used without making due allowance for the dependence of the Chézy coefficient on hydraulic mean depth, and hence on depth of the stream.)

Now $u_1 - u_w = (gD)^{0.5}$ (ref. 16 and 17)
So $u_w = u_1 - (gD)^{0.5}$
$= 2.098 - (9.81 \times 19.5 \times 10^{-2})^{0.5}$ m s$^{-1}$
$= +0.715$ m s$^{-1}$

Since this velocity is positive, small surface waves will move with a velocity of 0.715 m s$^{-1}$ in a downstream direction (consistent with shooting flow). The Froude number for the initial system (ref. 13, 16 and 18),

$$Fr = \frac{u}{(gD)^{0.5}}$$

$$= \frac{2.098}{(9.81 \times 19.5 \times 10^{-2})^{0.5}}$$

$$= \underline{1.52}$$

Finally, since $E = D + \dfrac{u^2}{2g}$ (ref. 11 and 12)

kinetic energy per unit weight of fluid,

$$\frac{u^2}{2g} = E - D$$

For the initial system, this kinetic energy

$$= 0.4193 - 19.5 \times 10^{-2} = 0.2243 \text{ m}$$

so the percentage of the specific energy which is due to kinetic energy in this system

$$= \frac{0.2243 \times 100}{0.4193}\%$$

$$= \underline{53.5\%}$$

In the tranquil system with fluid flowing at the alternative depth, kinetic energy per unit weight of fluid

$$= 0.4193 - 0.349 = 0.0703 \text{ m}$$

so the percentage of the specific energy due to kinetic energy

$$= \frac{0.0703 \times 100}{0.4193}\%$$

$$= \underline{16.8\%}$$

## 5.1.7 Alternative depths of streams and gradients of mild and steep slopes

*Problem*

Find the two alternative depths at which a stream of water (density = 998 kg m$^{-3}$) could flow in a channel of rectangular section (width = 40.0 cm) if the mass flow rate per unit width = 338 kg m$^{-1}$ s$^{-1}$ and specific energy for the system = 0.480 m. What would be the gradients of the mild and steep slopes required to maintain flow at the two alternative depths if the Manning roughness coefficient, $n = 0.015$ m$^{-0.333}$ s.

Acceleration due to gravity = 9.81 m s$^{-2}$

*Solution*

The two alternative depths for the water stream are the roots to the equation (ref. 11 and 12):

$$E = D + u^2/2g$$
$$= D + \frac{q^2}{2gD^2} \qquad (5.9)$$

In this case, volumetric flow rate per unit width of channel,

$$q = \frac{338}{998} = 0.3387 \text{ m}^2 \text{ s}^{-1}$$

So substituting data into equation (5.9)

$$0.480 = D + \frac{0.3387^2}{2 \times 9.81 \times D^2}$$

The roots to this equation may be obtained graphically, or by trial and error. Hence, the positive roots are:

$$D = 12.9 \text{ cm and } 45.1 \text{ cm}$$

i.e. the alternative depths at which water could flow in the channel = 12.9 cm and 45.1 cm

Now, the Manning formula (ref. 1, 2 and 3) indicates that

$$u = (1/n) m^{\frac{2}{3}} s^{\frac{1}{2}}$$

Since mass flow rate, $\qquad G = \rho BDu$

mass flow rate per unit width,

$$G/B = \rho Du$$
$$= \frac{\rho D}{n} m^{\frac{2}{3}} s^{\frac{1}{2}}$$

$$= \frac{\rho D}{n}\left(\frac{A}{P}\right)^{\frac{2}{3}} s^{\frac{1}{2}}$$

$$= \frac{\rho D}{n}\left(\frac{BD}{B+2D}\right)^{\frac{2}{3}} s^{\frac{1}{2}}$$

Now, the mild slope obtains by substituting $D = 45.1$ cm,

i.e. $$338 = \frac{998 \times 0.451}{0.015}\left(\frac{0.40 \times 0.451}{0.40 + 2 \times 0.451}\right)^{\frac{2}{3}} s^{\frac{1}{2}}$$

from which, slope $\quad s = 0.001\,770$ radians

$$= \underline{0.101 \text{ degrees}}$$

The steep slope obtains when $D = 12.9$ cm,

i.e. $$338 = \frac{998 \times 0.129}{0.015}\left(\frac{0.40 \times 0.129}{0.40 + 2 \times 0.129}\right)^{\frac{2}{3}} s^{\frac{1}{2}}$$

from which, slope $\quad s = 0.046\,21$ radians

$$= \underline{2.65 \text{ degrees}}$$

## 5.1.8 Critical flow conditions in channels of non-rectangular section

*Problem*

Calculate (i) the critical depth, (ii) the critical velocity, and (iii) the critical volumetric flow rate for liquid flowing in a channel of trapezoidal section when the specific energy of the stream $= 0.375$ m. The base of the channel $= 0.845$ m wide, and channel sides slope outwards at $60°$ to the horizontal.
Acceleration due to gravity $= 9.81$ m s$^{-2}$

*Solution*

(i) The critical velocity for fluid flowing in a channel of any-shaped cross section (ref. 19),

$$u_c = (g\bar{D}_c)^{0.5} \quad (5.10)$$

in which $\quad \bar{D}_c = A_c / B_{tc}$

$A_c$ and $B_{tc}$ can be calculated from the known proportions of the channel if critical depth of the stream, $D_c$, is known. $D_c$ for a channel of trapezoidal section may be calculated from (ref. 14):

$$5hD_c^2 + (3B - 4hE)D_c - 2EB = 0 \quad (5.11)$$

## 98 Problems in fluid flow

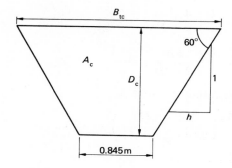

**Figure 5.4**

Now at critical depth, the stream section is as shown in Figure 5.4.

From this diagram, $\qquad 1/h = \tan 60°$

so $\qquad h = 0.5774$

Substituting in equation (5.11),

$$5 \times 0.5774 \times D_c^2 + (3 \times 0.845 - 4 \times 0.5774 \times 0.375)D_c - 2 \times 0.375 \times 0.845 = 0$$

i.e. $\qquad D_c^2 + 0.5781 D_c - 0.2195 = 0$

from which $\qquad D_c = 0.2615 \text{ m} \quad \text{or} \quad -0.8396 \text{ m}$

Since the positive value must be selected, critical depth of the stream,

$$D_c = \underline{26.15 \text{ cm}}$$

(ii) Now $\qquad A_c = D_c(0.845 + D_c \tan 30°) \text{ m}^2$

$\qquad\qquad\qquad = 0.2615(0.845 + 0.2615 \times 0.5774) \text{ m}^2$

$\qquad\qquad\qquad = 0.2605 \text{ m}^2$

Also, $\qquad B_{tc} = 0.845 + 2 \times D_c \tan 30° \text{ m}$

$\qquad\qquad\qquad = 0.845 + 2 \times 0.2615 \times 0.5774 \text{ m}$

$\qquad\qquad\qquad = 1.1470 \text{ m}$

So, $\qquad \bar{D}_c = \dfrac{0.2605}{1.1470} \text{ m}$

$\qquad\qquad\qquad = 0.2271 \text{ m}$

and using equation (5.10), critical velocity,

$$u_c = (9.81 \times 0.2271)^{0.5} \text{ m s}^{-1}$$

$$= \underline{1.493 \text{ m s}^{-1}}$$

(iii) Finally,

critical volumetric flow rate $= A_c u_c$

$$= 0.2605 \times 1.493 \text{ m}^3 \text{ s}^{-1}$$
$$= \underline{0.389 \text{ m}^3 \text{ s}^{-1}}$$

## 5.1.9 Flow measurement with broad-crested weirs

*Problem*

A flume in a channel of rectangular section consists of a contraction in width from $B_1$ to $B_2$, and at the neck, a long, flat topped hump of height $Z$ in the channel bed.

(i) If the liquid stream remains in tranquil flow while passing through the flume, and $D_1$ and $D_2$ represent stream depths well before and at the neck of the flume respectively, show that volumetric flow rate of the liquid will be given by:

$$Q = B_2 D_2 \left( \frac{2g(D_1 - D_2 - Z)}{1 - [(B_2 D_2)/(B_1 D_1)]^2} \right)^{0.5}$$

(ii) If the liquid stream is initially tranquil and slow moving, but accelerates to pass through critical conditions at the neck, before falling freely, or forming a standing wave downstream, show that, where S.I. units are used, volumetric flow rate of the stream will be given by:

$$Q = 1.705 B_2 (D_1 - Z)^{1.5} \text{ m}^3 \text{ s}^{-1}$$

(iii) Find the volumetric flow rate of liquid flowing in a channel of rectangular section if liquid streams in tranquil flow through a flat-bedded flume in the channel such that a 4.7% reduction in depth obtains from a normal depth of 1.27 m upstream due to a change in channel width from 2.95 m to 1.60 m at the neck.

(iv) If a standing wave forms downstream of a flat topped weir in a channel of rectangular section and uniform width $= 12.8$ m, find the volumetric flow rate of the stream if fluid depth well upstream of the weir is 2.58 m and the weir crest is 1.25 m above the bed of the channel.

Acceleration due to gravity $= 9.81$ m s$^{-2}$

*Solution*

(i) If the liquid stream remains in tranquil flow while passing through the flume, flow conditions may be represented as shown in Figure 5.5.
Since energy loss will be very small as liquid passes through the flume, we may

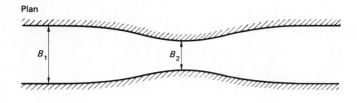

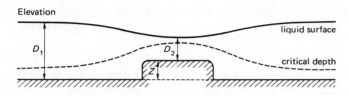

**Figure 5.5**

write:

$$E = D_1 + \frac{u_1^2}{2g} \quad \text{(ref. 11 and 12)}$$

$$= D_2 + Z + \frac{u_2^2}{2g}$$

If the two expressions on the right hand side are equated, on rearrangement, we have:

$$\frac{u_2^2}{2g} - \frac{u_1^2}{2g} = D_1 - D_2 - Z \tag{5.12}$$

But volumetric flow rate, $\quad Q = B_1 D_1 u_1$

$$= B_2 D_2 u_2$$

from which $\quad u_1 = \dfrac{B_2 D_2}{B_1 D_1} u_2$

Substituting this into equation (5.12)

$$\frac{u_2^2}{2g}\left[1 - \left(\frac{B_2 D_2}{B_1 D_1}\right)^2\right] = D_1 - D_2 - Z$$

which on rearrangement gives:

$$u_2 = \left(\frac{2g(D_1 - D_2 - Z)}{1 - [(B_2 D_2)/(B_1 D_1)]^2}\right)^{0.5}$$

But volumetric flow rate, when the liquid stream remains tranquil in the flume,

$$Q = B_2 D_2 u_2$$

So
$$Q = B_2 D_2 \left( \frac{2g(D_1 - D_2 - Z)}{1 - [(B_2 D_2)/(B_1 D_1)]^2} \right)^{0.5} \quad (5.13)$$

(ii) If an initially tranquil stream is accelerated such that its depth falls through critical as it passes through the flume, flow conditions may be represented as shown in Figure 5.6.

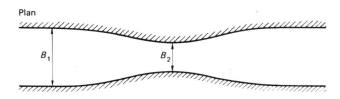

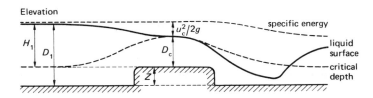

**Figure 5.6**

Taking the top of the hump as the energy datum level, we may write:

$$E = D_c + \frac{u_c^2}{2g} \quad \text{(ref. 11 and 12)}$$

from which, $u_c = [2g(E - D_c)]^{0.5}$

But $Q = B_2 D_c u_c$

$\phantom{Q} = B_2 D_c [2g(E - D_c)]^{0.5}$

Now since the channel has a rectangular section,

$$D_c = \frac{2}{3} E \quad \text{(ref. 13, 14 and 15)}$$

So $Q = \frac{2}{3} B_2 E [2g(E - \frac{2}{3}E)]^{0.5}$

which, if S.I. units are used, simplifies to (cf. ref. 20 and 21):

$$Q = 1.705 \, B_2 E^{1.5} \text{ m}^3 \text{ s}^{-1} \quad (5.14)$$

But since there will be little energy loss upstream of the weir, we may also write:

$$E = H_1 + \frac{u_1^2}{2g}$$

So, if the stream upstream is slow moving,

$$E \approx H_1$$

and since

$$H_1 = D_1 - Z$$

equation (5.14), which obtains when stream depth falls through critical in the flume, becomes (cf. ref. 20):

$$Q = 1.705 \, B_2 (D_1 - Z)^{1.5} \text{ m}^3 \text{ s}^{-1} \tag{5.15}$$

(iii) Now where the flow remains tranquil as the stream passes through the flume, using equation (5.13), volumetric flow rate

$$Q = 1.60(1 - 0.047) \times 1.27 \times \left( \frac{2 \times 9.81 \times 0.047 \times 1.27}{1 - \left( \frac{1.60(1 - 0.047) \times 1.27}{2.95 \times 1.27} \right)^2} \right)^{0.5} \text{ m}^3 \text{ s}^{-1}$$

$$\underline{= 2.45 \text{ m}^3 \text{ s}^{-1}}$$

(iv) Where a standing wave forms downstream of a flat topped weir in a channel of uniform, rectangular section, we find, using equation (5.15), that volumetric flow rate,

$$Q = 1.705 \times 12.8 \, (2.58 - 1.25)^{1.5} \text{ m}^3 \text{ s}^{-1}$$
$$= 33.5 \text{ m}^3 \text{ s}^{-1}$$

## 5.1.10 Gradually varied flow behind a weir

*Problem*

A straight channel of rectangular section (bed slope = 0.15 degrees, width = 5.75 m) is carrying water at a volumetric flow rate of $16.8 \text{ m}^3 \text{ s}^{-1}$.
(i) Find the normal depth at which the stream would flow if the Manning roughness coefficient, $n = 0.022 \text{ m}^{-0.333}$ s, and
(ii) if there is a weir in the channel to make fluid depth there = 2.40 m, calculate the distance upstream from the weir at which stream depth = 1.50 m.
Acceleration due to gravity = $9.81 \text{ m s}^{-2}$

*Solution*

(i) The Manning formula gives average fluid velocity (ref. 1, 2 and 3),

$$u = \left( \frac{1}{n} \right) m^{\frac{2}{3}} s^{\frac{1}{2}}$$

so volumetric flow rate,

$$Q = \frac{BD_n}{n} \, m^{\frac{2}{3}} s^{\frac{1}{2}}$$

$$= \frac{BD_n}{n}\left(\frac{BD_n}{B+2D_n}\right)^{\frac{2}{3}} s^{\frac{1}{2}} \qquad (5.16)$$

Now bed slope,
$$s = 0.15 \text{ degrees}$$

$$= \frac{0.15 \times \pi}{180} \text{ radians}$$

$$= 0.002\,618 \text{ radians}$$

So substituting data into equation (5.16)

$$16.8 = \frac{5.75 \times D_n}{0.022}\left(\frac{5.75 \times D_n}{5.75 + 2 \times D_n}\right)^{\frac{2}{3}} \times 0.002\,618^{0.5}$$

This equation may be solved graphically, or by trial and error. Hence, the normal depth of the stream,

$$\underline{D_n = 1.336 \text{ m}}$$

(ii) Now the rate of change of stream depth with distance along the channel is given by (ref. 22 and 23):

$$\frac{dD}{dl} = s \cdot \frac{1-(D_n/D)^c}{1-(D_c/D)^3} \qquad (5.17)$$

in which
$$c = \frac{(1+a)}{b}$$

and $a$ and $b$ are indices in the Manning resistance formula,

$$u = \left(\frac{1}{n}\right) m^a s^b$$

Thus,
$$c = \frac{(1+\frac{2}{3})}{\frac{1}{2}}$$

$$= 3.333$$

Now in rectangular channels, volumetric flow rate per unit width, (ref. 13 and 15)

$$q = (gD_c^3)^{0.5}$$

104    *Problems in fluid flow*

so critical depth of the stream,

$$D_c = \left(\frac{q^2}{g}\right)^{\frac{1}{3}}$$

$$= \left(\frac{Q^2}{gB^2}\right)^{\frac{1}{3}}$$

$$= \left(\frac{16.8^2}{9.81 \times 5.75^2}\right)^{\frac{1}{3}} \text{ m}$$

$$= 0.9547 \text{ m}$$

Rearranging equation (5.17), for intervals of depth $\delta D$,

$$\delta l = \frac{\delta D}{s} \cdot \frac{1 - (D_c/D)^3}{1 - (D_n/D)^c}$$

so substituting data,

$$\delta l = \frac{\delta D}{0.002\,618} \left(\frac{1 - (0.9547/D)^3}{1 - (1.336/D)^{3.333}}\right) \quad (5.18)$$

Let us now divide the depth range 2.40 m to 1.50 m into intervals, $\delta D = 0.10$ m. Starting from known conditions at the weir, we may calculate mean depth for each interval and substitute these mean values for $D$ into equation (5.18) to calculate the length interval, $\delta l$, corresponding to each depth interval of 0.10 m. Data obtained in this way are given in Table 5.1.

**Table 5.1**

| Depth Range (m) | Mean Depth (m) | $\delta l$ (m) |
|---|---|---|
| 2.40–2.30 | 2.35 | 42.04 |
| 2.30–2.20 | 2.25 | 42.81 |
| 2.20–2.10 | 2.15 | 43.83 |
| 2.10–2.00 | 2.05 | 45.18 |
| 2.00–1.90 | 1.95 | 47.06 |
| 1.90–1.80 | 1.85 | 49.76 |
| 1.80–1.70 | 1.75 | 53.93 |
| 1.70–1.60 | 1.65 | 60.96 |
| 1.60–1.50 | 1.55 | 74.95 |

Hence, by summing all the length intervals, $\delta l$, the distance upstream from the weir at which stream depth is 1.50 m = 460.5 m

## 5.1.11 Analysis of a hydraulic jump

*Problem*

A wide channel of rectangular section, carrying water at a volumetric flow rate per unit width of 1.49 m² s⁻¹, changes slope from 1.85 degrees to 0.035 degrees. Calculate:
(i) the critical depth of the stream,
(ii) the normal depths for the stream on both upstream and downstream slopes, and hence show that a hydraulic jump can be expected to occur in the system (Manning roughness coefficient, $n = 0.021$ m$^{-0.333}$ s),
(iii) the conjugate depths of fluid on upstream and downstream slopes, and hence establish on which slope the jump will occur,
(iv) how far away the jump will occur from the location at which the slope changes (a) using an accurate method of calculation, and (b) assuming that the total energy gradient line is linear over this region,
(v) the power loss by fluid in the jump per unit width of channel.

Acceleration due to gravity = 9.81 m s⁻²
Density of water = 998 kg m⁻³

*Solution*

(i) Volumetric flow rate per unit width (ref. 13 and 15),

$$q = (gD_c^3)^{0.5}$$

so critical depth,

$$D_c = \left(\frac{q^2}{g}\right)^{\frac{1}{3}}$$

$$= \left(\frac{1.49^2}{9.81}\right)^{\frac{1}{3}} \text{ m}$$

$$= \underline{0.609 \text{ m}}$$

(ii) Now, $\quad q = \dfrac{D_n}{n} m^{\frac{2}{3}} s^{\frac{1}{2}} \quad$ (ref. 1, 2 and 3)

which for wide channels, $\quad \approx \dfrac{D_n^{\frac{5}{3}} s^{\frac{1}{2}}}{n}$

so that $\quad D_n = \left(\dfrac{nq}{s^{0.5}}\right)^{3/5}$

Since the upstream slope, $s = 1.85$ degrees $= 0.032\,29$ radians the normal depth for flow over it,

$$D_{n,u} = \left(\frac{0.021 \times 1.49}{0.032\,29^{0.5}}\right)^{3/5} \text{ m}$$

$$= \underline{0.350 \text{ m}}$$

The downstream slope,     $s = 0.035$ degrees

$$= 0.000\,610\,9 \text{ radians}$$

So the normal depth for flow over it,

$$D_{n,d} = \left(\frac{0.021 \times 1.49}{0.000\,610\,9^{0.5}}\right)^{\frac{3}{5}} \text{ m}$$

$$= \underline{1.152\,\text{m}}$$

Since $D_{n,u} < D_c < D_{n,d}$, a hydraulic jump may be expected to occur somewhere in the system.

(iii) $D_{n,u} < D_c$, so setting $D_{n,u} = D_1$, the conjugate depth for a tranquil stream on the upstream slope will be given by (ref. 24 and 25):

$$D_{2,u} = -\tfrac{1}{2}D_1\left[1 - \left(1 + \frac{8u_1^2}{gD_1}\right)^{0.5}\right]$$

$$= -\tfrac{1}{2}D_1\left[1 - \left(1 + \frac{8q^2}{gD_1^3}\right)^{0.5}\right]$$

Substituting data,    $D_{2,u} = -\dfrac{0.350}{2}\left[1 - \left(1 + \dfrac{8 \times 1.49^2}{9.81 \times 0.350^3}\right)^{0.5}\right]$ m

$$= \underline{0.976\,\text{m}}$$

$D_{n,d} > D_c$, so setting $D_{n,d} = D_2$, the conjugate depth for a shooting stream on the downstream slope will be given by (ref. 26):

$$D_{1,d} = -\tfrac{1}{2}D_2\left[1 - \left(1 + \frac{8q^2}{gD_2^3}\right)^{0.5}\right]$$

Substituting data,    $D_{1,d} = -\dfrac{1.152}{2}\left[1 - \left(1 + \dfrac{8 \times 1.49^2}{9.81 \times 1.152^3}\right)^{0.5}\right]$ m

$$= \underline{0.275\,\text{m}}$$

However, conjugate depth $D_{2,u}$ is less than normal depth $D_{n,d}$, so the hydraulic jump will occur on the upstream slope, and this will be followed by a smooth transition in tranquil flow to the normal depth of 1.152 m on the downstream slope.

(iv) (a) Slope of the liquid surface with respect to the channel bed, (ref. 22 and 23),

$$\frac{dD}{dl} = s\left(\frac{1 - (D_n/D)^c}{1 - (D_c/D)^3}\right) \tag{5.19}$$

in which     $c = (1 + a)/b$

and $a$ and $b$ are indices in the Manning formula,

$$u = (1/n)m^a s^b$$

Now
$$c = \frac{(1+\frac{2}{3})}{\frac{1}{2}}$$
$$= 3.333$$

So by rearranging equation (5.19), a depth change, $\delta D$, occurs over a distance,

$$\delta l = \frac{\delta D}{s}\left(\frac{1-(D_c/D)^3}{1-(D_n/D)^{3.333}}\right) \quad (5.20)$$

If it may be assumed that normal depth for the downstream slope is attained where bed slope changes, a depth change from 0.976 m to 1.152 m must take place on the upstream slope for which $D_n = 0.350$ m.

Let us divide the overall depth change into 8 intervals, $\delta D = 0.022$ m. Then if mean depth for each interval is calculated and substituted as $D$ in equation (5.20), length interval, $\delta l$, may be computed for each step. Thus, substituting data,

$$\delta l = \frac{0.022}{0.032\,29}\left(\frac{1-(0.609/D)^3}{1-(0.350/D)^{3.333}}\right) \text{m}$$

Data calculated from this equation are shown in Table 5.2.

**Table 5.2**

| Depth Range (m) | Mean Depth (m) | $\delta l$ (m) |
|---|---|---|
| 1.152–1.130 | 1.141 | 0.589 |
| 1.130–1.108 | 1.119 | 0.584 |
| 1.108–1.086 | 1.097 | 0.578 |
| 1.086–1.064 | 1.075 | 0.571 |
| 1.064–1.042 | 1.053 | 0.564 |
| 1.042–1.020 | 1.031 | 0.556 |
| 1.020–0.998 | 1.009 | 0.548 |
| 0.998–0.976 | 0.987 | 0.538 |

Hence, by adding the length intervals, $\delta l$, we find that the hydraulic jump will occur 4.53 m on the upstream side of the slope change.

(b) If it is assumed that the total energy line has a constant gradient, $i$, over this interval, the distance between hydraulic jump and slope change location may be given by (ref. 27):

$$l = \frac{E_2 - E_1}{s - i} \quad (5.21)$$

in which $E_1$ = specific energy on the downstream side of the hydraulic jump,
$E_2$ = specific energy where slope changes,
$s$ = slope of the channel bed,
$i$ = slope of the total energy line.

(Since for normal flow, $i = s$, $i$ may be calculated from the Manning formula assuming average values for fluid velocity and hydraulic mean depth.)

Now,
$$E = D + \frac{u^2}{2g} \quad \text{(ref. 11 and 12)}$$

$$= D + \frac{q^2}{2gD^2}$$

So,
$$E_1 = 0.976 + \frac{1.49^2}{2 \times 9.81 \times 0.976^2} \text{ m}$$
$$= 1.095 \text{ m}$$

and
$$E_2 = 1.152 + \frac{1.49^2}{2 \times 9.81 \times 1.152^2} \text{ m}$$
$$= 1.237 \text{ m}$$

Since the channel is wide, average hydraulic mean depth may be taken

$$= \frac{1.152 + 0.976}{2} = 1.064 \text{ m}$$

Average fluid velocity
$$= \frac{1}{2}\left(\frac{q}{D_1} + \frac{q}{D_2}\right)$$

$$= \frac{1.49/1.152 + 1.49/0.976}{2} \text{ m s}^{-1}$$

$$= 1.410 \text{ m s}^{-1}$$

So using the Manning formula,

$$1.410 = \frac{1.064^{\frac{2}{3}} \times i^{\frac{1}{2}}}{0.021}$$

from which
$$i = 0.000\,807\,2 \text{ radians}$$

Finally, substituting data into equation (5.21), the approximate distance between hydraulic jump and the place where bed slope changes,

$$l = \frac{1.237 - 1.095}{0.032\,29 - 0.000\,807\,2} \text{ m}$$

$$= \underline{4.510 \text{ m}}$$

(v) If $E_u$ = specific energy of the stream immediately upstream of the jump, and $E_d$ = specific energy of the stream immediately after the jump, power loss per unit width of channel (ref. 28)

$$= \rho q g (E_u - E_d)$$

In this case, we have just found

$$E_d = E_1 \text{ in section (iv)},$$
$$= 1.095 \text{ m}$$

Also,
$$E_u = 0.350 + \frac{1.49^2}{2 \times 9.81 \times 0.350^2} \text{ m}$$
$$= 1.274 \text{ m}$$

So power loss per unit width of stream in the hydraulic jump
$$= 998 \times 1.49 \times 9.81(1.274 - 1.095) \text{ W}$$
$$= \underline{2.61 \text{ kW}}$$

## 5.2 Student Exercises

1  Use the Manning formula to calculate volumetric flow rate of water if it forms a stream 54.5 cm deep and 1.65 m wide while flowing through a rough bedded channel of rectangular section, inclined downwards at an angle of 0.19 degrees to the horizontal. Find also the shear stress which would be experienced by the channel walls.

Manning roughness coefficient, $n = 0.029 \text{ m}^{-0.333} \text{ s}$
Density of water $= 998 \text{ kg m}^{-3}$
Acceleration due to gravity $= 9.81 \text{ m s}^{-2}$

2  Use the Manning formula to find the depth of a fluid stream flowing in a straight, wood-plank channel of rectangular section when the volumetric flow rate $= 185 \text{ dm}^3 \text{ s}^{-1}$, width of the channel $= 74 \text{ cm}$, slope of the channel bed $= 0.45$ degrees, and the Manning roughness coefficient, $n$, for the material of construction $= 0.012 \text{ m}^{-0.333} \text{ s}$. If the Chézy formula were to be used to describe flow in the system, find the value which would obtain for the Chézy coefficient, $C$.

3  (i) For a given bed slope, find the fluid depth at which volumetric flow rate will be a maximum in a channel of trapezoidal cross section if cross-sectional

area of the fluid stream must $= 2.16\,\text{m}^2$ and the channel sides slope at $25°$ to the horizontal.

(ii) Find the slope at which this channel must be laid to carry water at $1.80\,\text{m}^3\,\text{s}^{-1}$. The Manning roughness coefficient, $n$, may be taken $= 0.020\,\text{m}^{-0.333}\,\text{s}$.

4  (i) Using the Manning formula, find how deep a stream will be if it flows with a free surface in a closed conduit of circular cross section (2.8 m diameter) at maximum volumetric flow rate.

(ii) What will be the volumetric flow rate of the stream then if the channel has a slope of 0.035 degrees, and the Manning roughness coefficient, $n = 0.016\,\text{m}^{-0.333}\,\text{s}$?

5  (i) If upstream velocity is small enough to be neglected, calculate the volumetric flow rate of a stream if fluid pouring freely through a rectangular notch (85.0 cm wide) is 42.6 cm deep in the notch (a) if no allowance is made for end contractions at the weir, and (b) if the Francis formula is used to allow for the notch being narrower than the main channel and is situated symmetrically across the end of it.

(ii) Find the depth of a stream at a rectangular notch (55.0 cm wide) if the stream pours freely through the notch at $385\,\text{dm}^3\,\text{s}^{-1}$ (a) if no allowance is made for end contractions at the weir, and (b) if the Francis formula is used to allow for the notch being narrower than the main channel and situated asymmetrically across the end of it, with one edge flush with the side of the main channel.

(iii) Calculate the volumetric flow rate of a stream if liquid pours freely through a V-shaped notch (base angle $= 60°$) in a sharp-edged weir and has a depth of 24.6 cm at the centre of the notch.

Coefficient of discharge in all cases above $= 0.61$

Acceleration due to gravity $= 9.81\,\text{m s}^{-2}$

6  (i) What will be the specific energy of a stream of fluid flowing in a channel of rectangular section if the stream is 2.75 m wide, 87.4 cm deep, and flowing at $1.15\,\text{m}^3\,\text{s}^{-1}$?

(ii) Calculate the Froude number for the system in part (i).

(iii) What will be the alternative depth at which fluid could flow with the same volumetric flow rate and specific energy as the stream in part (i)?

(iv) Calculate the Froude number for the stream in part (iii).

(v) Find the percentage of the specific energy which is due to kinetic energy for the system in part (i).

(vi) Calculate the percentage of the specific energy which is due to kinetic energy for the system in part (iii).

(vii) Find the maximum volumetric flow rate which could obtain by suitable selection of bed slope at the same specific energy as the streams of parts (i) and (iii).

Acceleration due to gravity = $9.81 \, \mathrm{m\,s^{-2}}$

7  Consider a stream of specific energy = 0.685 m, flowing at a volumetric flow rate of $1.07 \, \mathrm{m^3\,s^{-1}}$ in a channel of rectangular section and width 1.25 m. Find the two alternative depths at which fluid could flow and the values of the mild and steep slopes of the channel bed which would be required to maintain flow at these two depths.

Manning roughness coefficient, $n = 0.018 \, \mathrm{m^{-0.333}\,s}$

Acceleration due to gravity = $9.81 \, \mathrm{m\,s^{-2}}$

8  Show that the critical depth for fluid flowing in a V-shaped channel = 0.8 × specific energy of the stream.

Calculate (i) critical depth, (ii) critical velocity and (iii) maximum volumetric flow rate for fluid flowing in a V-shaped channel of base angle = 65° when specific energy of the stream = 0.655 m.

Acceleration due to gravity = $9.81 \, \mathrm{m\,s^{-2}}$

9  (i) A stream flows at $5.42 \, \mathrm{m^3\,s^{-1}}$ through a channel of rectangular section and constant width (= 3.85 m). If there is a broad-crested weir in the channel, rising 44 cm above bed level, and fluid flows over it at a depth of 62.7 cm, establish that the stream will remain in tranquil flow as it flows over the weir, then find how deep the stream will be just before the liquid surface starts to dip to flow over the weir.

(ii) To what height would the weir have to be adjusted to just cause a standing wave to form downstream at the same flow rate if, as a first approximation, this were considered to have no effect on the stream depth upstream?

Acceleration due to gravity = $9.81 \, \mathrm{m\,s^{-2}}$

10  Water flows at $0.368 \, \mathrm{m^3\,s^{-1}}$ through a channel of rectangular section and width = 1.15 m, and in so doing, passes through three reaches (a, b and c) in which the bed slopes successively at 3.25°, 0.75° and 0.013°. If each reach is long enough for steady, uniform flow to be attained in it, establish:
(i) that, and on which slope a hydraulic jump will occur,
(ii) the exact location of the jump, and
(iii) the power loss by fluid in the jump.

Manning roughness coefficient, $n = 0.019 \, \mathrm{m^{-0.333}\,s}$

Density of water = $998 \, \mathrm{kg\,m^{-3}}$

Acceleration due to gravity = $9.81 \, \mathrm{m\,s^{-2}}$

112   *Problems in fluid flow*

## 5.3 References

References are abbreviated as follows:

D, G & S = J. F. Douglas, J. M. Gasiorek and J. A. Swaffield, Fluid Mechanics, Pitman, 1980

F & M = J. R. D. Francis and P. Minton, Civil Engineering Hydraulics, Edward Arnold, 5th edition, 1984

1. D, G & S, page 436
2. F & M, page 301
3. F & M, page 233
4. D, G & S, page 433
5. J. M. Coulson and J. F. Richardson, Chemical Engineering, Vol. 1, 3rd edition, Pergamon, 1977, page 63
6. D, G & S, page 11
7. F & M, page 5
8. D, G & S, page 178
9. J. F. Douglas, Solution of Problems in Fluid Mechanics, Part 1, Pitman, 1975, page 116
10. D, G & S, page 179
11. D, G & S, page 449
12. F & M, page 273
13. D, G & S, page 450
14. D, G & S, page 453
15. F & M, page 277
16. D, G & S, page 151
17. F & M, page 320
18. F & M, page 146
19. D, G & S, page 452
20. D, G & S, page 458
21. F & M, page 287
22. D, G & S, page 462
23. F & M, page 309
24. D, G & S, page 466
25. F & M, page 292
26. D, G & S, page 467
27. D, G & S, page 469
28. D, G & S, page 470

## 5.4 Notation

| Symbol | Description | Unit |
|---|---|---|
| $a$ | power to which $m$ (hydraulic mean depth) is raised in the Manning formula | |
| $A$ | cross-sectional area of the stream | $m^2$ |
| $A_c$ | cross-sectional area of a fluid stream under critical (maximum flow rate) conditions | $m^2$ |
| $b$ | power to which $s$ (bed slope) is raised in the Manning formula | |
| $B$ | width of a channel of rectangular section | m |
| $B_{tc}$ | width of the free surface of a stream under critical (maximum flow rate) conditions | m |
| $B_1$ | width of a channel of rectangular section at section 1 (upstream) | m |
| $B_2$ | width of a channel of rectangular section at section 2 (downstream) | m |
| $c$ | $(1+a)/b$ | |
| $C$ | Chézy coefficient | $m^{0.5} s^{-1}$ |
| $C_D$ | coefficient of discharge | |
| $D$ | depth of stream | m |
| $D_c$ | critical depth of a fluid stream | m |
| $D_n$ | normal depth of a stream | m |

Flow and flow measurement in open channels 113

| Symbol | Description | Unit |
|---|---|---|
| $\bar{D}_c$ | the ratio $A_c/B_{tc}$ | m |
| $D_{n,d}$ | normal depth of a stream on a downstream slope | m |
| $D_{n,u}$ | normal depth of a stream on an upstream slope | m |
| $D_1$ | depth of a stream at section 1 (upstream) | m |
| $D_2$ | depth of a stream at section 2 (downstream) | m |
| $D_{1,d}$ | conjugate depth for a shooting stream on a downstream slope | m |
| $D_{2,u}$ | conjugate depth for a tranquil stream on an upstream slope | m |
| $E$ | specific energy of a fluid stream | m |
| $E_d$ | specific energy of a stream immediately after a hydraulic jump | m |
| $E_u$ | specific energy of a stream immediately upstream of a hydraulic jump | m |
| $E_1$ | specific energy of a stream on the downstream side of a hydraulic jump | m |
| $E_2$ | specific energy of a stream where bed slope changes | m |
| $Fr$ | Froude number | |
| $g$ | acceleration due to gravity | $m\,s^{-2}$ |
| $G$ | mass flow rate | $kg\,s^{-1}$ |
| $h$ | distance in a horizontal direction over which a channel side slopes for unit increase in height (see Figures 5.2 and 5.4) | m |
| $h'$ | distance below top free surface of a liquid stream | m |
| $H$ | depth of fluid in a stream with the top of a weir being taken as datum level | m |
| $H_1$ | depth of fluid in a stream at section 1 (upstream) with top of a weir being taken as datum level | m |
| $i$ | gradient of the total energy line | radians |
| $k$ | length of wetted side of a channel (see Figure 5.1) | m |
| $l$ | distance along a channel | m |
| $m$ | hydraulic mean depth ($= A/P$) of a stream | m |
| $n$ | Manning resistance coefficient | $m^{-0.333}\,s$ |
| $n'$ | number of end contractions at a notch in a weir | |
| $P$ | length of wetted perimeter in a cross section | m |
| $q$ | volumetric flow rate per unit width | $m^2\,s^{-1}$ |
| $q_{max}$ | volumetric flow rate per unit width of a stream at critical (maximum flow) conditions | $m^2\,s^{-1}$ |
| $Q$ | volumetric flow rate | $m^3\,s^{-1}$ |
| $r$ | radius of a conduit of circular cross section | m |
| $R_m$ | shear stress experienced by channel walls | $N\,m^{-2}$ |
| $s$ | slope of the channel bed | radians |
| $u$ | average velocity of fluid in a stream | $m\,s^{-1}$ |
| $u_c$ | average fluid velocity under critical (maximum flow rate) conditions | $m\,s^{-1}$ |
| $u_w$ | velocity at which a small wave travels relative to the channel walls | $m\,s^{-1}$ |

| Symbol | Description | Unit |
|---|---|---|
| $u_1$ | average fluid velocity of a stream at section 1 (upstream) | $m\,s^{-1}$ |
| $u_2$ | average fluid velocity of a stream at section 2 (downstream) | $m\,s^{-1}$ |
| $W$ | width of the free surface of a stream | m |
| $y$ | distance from channel walls | m |
| $Z$ | height of a broad-crested weir above bed level | m |
| $\delta D$ | an increment in fluid depth occurring over distance increment, $\delta l$ | m |
| $\delta l$ | an increment in distance along a channel | m |
| $\theta$ | half of the angle subtended by a free surface of liquid at the centre of a pipeline of circular cross section | radians |
| $\theta'$ | half of the base angle of a V-shaped channel | radians |
| $\rho$ | density of fluid | $kg\,m^{-3}$ |
| $\mu$ | viscosity of the fluid | $kg\,m^{-1}\,s^{-1}$ |

# 6 Pumping of liquids

## 6.1 Worked Examples

### 6.1.1 Cavitation and its avoidance in suction pipes

*Problem*

A centrifugal pump is situated 12.5 m vertically above the surface of liquid toluene in a tank which is open to the atmosphere, and at 30 °C. Show that, even if primed, the pump could not lift liquid toluene from the tank under these conditions.

If the minimum net positive suction head of the pump = 0.650 m of water under these conditions, and the inside diameter of the suction pipe = 2.30 cm, calculate:
(i) the maximum height at which the pump could be located above the toluene surface to deliver the liquid at 1.80 dm$^3$ s$^{-1}$ after priming, without risk of cavitation, and if the suction pipe just dipped into toluene in the tank (the pipe may be considered smooth),
(ii) the maximum delivery rate (in dm$^3$ s$^{-1}$) after priming which could be obtained without risk of cavitation if the pump were located 9.00 m above the liquid surface with suction pipe just dipping into the liquid [friction factor $\phi$ may be considered to have the same value as in section (i)], and
(iii) the minimum inside diameter (in cm) of suction pipe which could permit delivery at 1.80 dm$^3$ s$^{-1}$ after priming with the pump located 9.00 m above the liquid surface. (The pipe may again be considered smooth.)

Atmospheric pressure on the day = 100.2 kPa
Acceleration due to gravity = 9.81 m s$^{-2}$
Density of water at 30 °C = 996 kg m$^{-3}$
Density of liquid toluene = 867 kg m$^{-3}$
Saturated vapour pressure of toluene at 30 °C = 4.535 kPa
Viscosity of liquid toluene at 30 °C = 5.26 × 10$^{-4}$ kg m$^{-1}$ s$^{-1}$
For turbulent flow in smooth pipes, $\phi = 0.0396\, Re^{-0.25}$

## Solution

The maximum height of a column of toluene which could be supported by atmospheric pressure on the day

$$= \frac{(100.2 - 4.535) \times 10^3}{867 \times 9.81} \text{ m}$$

$$= 11.25 \text{ m}$$

So with a pump situated 12.5 m above the liquid surface in the tank, pumping would merely cause flash vaporisation of toluene from the top of the column of liquid.

(i) Now the net positive suction head of the pump = 0.650 m of water
So the net positive suction head of the pump

$$= 0.650 \times \frac{996}{867} = 0.747 \text{ m of toluene}$$

Since in this case, with the reservoir of liquid located below the pump (ref. 1, 2 and 3)

$$\text{N.P.S.H.} = \left(\frac{P_1}{\rho g} - z_i - h_f\right) - \frac{P_{\text{liq}}}{\rho g}$$

so that

$$z_i = \frac{P_1}{\rho g} - \frac{P_{\text{liq}}}{\rho g} - \text{N.P.S.H.} - \frac{4\phi l u^2}{dg} \quad (6.1)$$

(ref. 4 and 5)

Since the suction pipe is vertical and just dips into liquid in the tank,

$$l = z_i \quad (6.2)$$

Also, volumetric flow rate,

$$Q = \frac{\pi d^2 u}{4}$$

so that

$$u = \frac{4Q}{\pi d^2} \quad (6.3)$$

Finally, the Reynolds number for fluid flow,

$$Re = \frac{\rho u d}{\mu}$$

$$= \frac{4Q\rho}{\pi d \mu}$$

$$= \frac{4 \times 1.80 \times 10^{-3} \times 867}{\pi \times 2.30 \times 10^{-2} \times 5.26 \times 10^{-4}}$$

$$= 1.642 \times 10^5$$

Thus, flow will be turbulent in the suction pipe

so that, 
$$\phi = 0.0396 \, Re^{-0.25}$$
$$= 1.967 \times 10^{-3} \tag{6.4}$$

So substituting from equations (6.2), (6.3) and (6.4) and other data also into equation (6.1)

$$z_i = \frac{100.2 \times 10^3}{867 \times 9.81} - \frac{4.535 \times 10^3}{867 \times 9.81} - 0.747$$
$$- \frac{4 \times 1.967 \times 10^{-3} \times 4^2 \times Q^2}{\pi^2 d^5 g} z_i$$

$$= 10.501 - \frac{4 \times 1.967 \times 10^{-3} \times 16 \, (1.80 \times 10^{-3})^2}{\pi^2 \, (2.30 \times 10^{-2})^5 \times 9.81} z_i$$

$$= 10.501 - 0.6545 z_i$$

so that, 
$$z_i = \frac{10.501}{1.6545} \, \text{m}$$
$$= 6.347 \, \text{m}$$

Thus the maximum height above toluene in the tank that the pump could be located without risk of cavitation while pumping liquid at $1.80 \, \text{dm}^3 \, \text{s}^{-1}$ = 6.347 m.

(ii) 
$$\text{N.P.S.H.} = \frac{P_1}{\rho g} - \frac{P_{\text{liq}}}{\rho g} - z_i - \frac{4\phi l u^2}{dg} \quad [\text{see section (i)}]$$

Thus, 
$$\frac{4\phi l u^2}{dg} = \frac{100.2 \times 10^3}{867 \times 9.81} - \frac{4.535 \times 10^3}{867 \times 9.81} - 9.00 - 0.747$$

i.e. 
$$\frac{4 \times 1.967 \times 10^{-3} \times 9.00 \, u^2}{2.30 \times 10^{-2} \times 9.81} = 1.501$$

from which, 
$$u = 2.187 \, \text{m s}^{-1}$$

But, 
$$Q = \frac{\pi d^2 u}{4}$$
$$= \frac{\pi (2.30 \times 10^{-2})^2 \times 2.187 \times 10^3}{4} \, \text{dm}^3 \, \text{s}^{-1}$$
$$= \underline{0.909 \, \text{dm}^3 \, \text{s}^{-1}}$$

Thus, the maximum delivery rate if the pump were located 9.00 m above toluene in the tank = $0.909 \, \text{dm}^3 \, \text{s}^{-1}$.

(iii) Now again,
$$\text{N.P.S.H.} = \frac{P_1}{\rho g} - \frac{P_{\text{liq}}}{\rho g} - z_i - \frac{4\phi l u^2}{dg} [\text{see section (i)}] \tag{6.5}$$

118  *Problems in fluid flow*

This time, fluid velocity would depend on the pipe diameter to be selected since

$$u = \frac{4Q}{\pi d^2} \tag{6.6}$$

Also, $\phi$ would depend on pipe diameter,

since
$$\phi = 0.0396\, Re^{-0.25} \tag{6.7}$$

and
$$Re = \frac{\rho u d}{\mu}$$

$$= \frac{4Q\rho}{\pi \mu d} \tag{6.8}$$

So rearranging equation (6.5),

$$\frac{4\phi l u^2}{dg} = \frac{P_1}{\rho g} - \frac{P_{liq}}{\rho g} - z_i - N.P.S.H.$$

and substituting from equations (6.6), (6.7) and (6.8), and introducing further data,

$$4 \times 0.0396 \left(\frac{4Q\rho}{\pi \mu d}\right)^{-0.25} \times \frac{9.00 \times 16\, Q^2}{\pi^2 d^5 g} = \frac{100.2 \times 10^3}{867 \times 9.81}$$

$$- \frac{4.535 \times 10^3}{867 \times 9.81} - 9.00 - 0.747$$

so that,

$$0.2356 \left(\frac{4 \times 1.80 \times 10^{-3} \times 867}{\pi \times 5.26 \times 10^{-4}}\right)^{-0.25} \times \frac{(1.80 \times 10^{-3})^2}{d^{4.75}} = 1.501$$

from which,     $d^{4.75} = 6.487 \times 10^{-8}$

and     $d = 3.07$ cm

i.e. the minimum inside diameter of smooth suction pipe which could permit delivery of 1.80 dm³ s⁻¹ of toluene while the pump is located 9.00 m above the liquid surface = 3.07 cm.

## 6.1.2  Specific speed of a centrifugal pump, and similarity in centrifugal pump systems

*Problem*

A centrifugal pump (with outside diameter of its impeller = 11.8 cm) operates, while delivering water at 24.8 dm³ s⁻¹, against a fluid head of 14.7 m. If the pump's impeller is then turning at 1450 rev min$^{-1}$, evaluate its characteristic specific speed, $N_s$.

A geometrically similar pump is now required to deliver water at 48.0 dm$^3$ s$^{-1}$ against a system head which is expected to be 15% greater than that above. For dynamical similarity to obtain as well, how fast must the new pump's impeller turn to do this, and what must be the diameter of the new pump's impeller?

*Solution*

The specific speed of a pump may be defined thus (cf. ref. 6 and 7):

$$N_s = \frac{NQ^{0.5}}{H^{0.75}}$$

(Notice that when defined this way, specific speed is not dimensionless.) For the first pump,

$$N_s = \frac{1450\,(24.8 \times 10^{-3})^{0.5}}{60(14.7)^{0.75}} \text{ m}^{0.75}\text{ s}^{-1.5}$$

$$= 0.507 \text{ m}^{0.75}\text{ s}^{-1.5}$$

The new pump system is to be geometrically and dynamically similar, and deliver water at 48.0 dm$^3$ s$^{-1}$ against a head of liquid

$$= 14.7 \times \frac{115}{100} = 16.9 \text{ m}$$

Thus, the new pump's turning rate,

$$N_2 = \frac{0.507 \times 16.9^{0.75} \times 60}{(48.0 \times 10^{-3})^{0.5}} \text{ rev min}^{-1}$$

$$= \underline{1157 \text{ rev min}^{-1}}$$

Delivery head, and size and speed of a centrifugal pump together give rise to a dimensionless group,

$$\pi_4 = \frac{gH}{D^2 N^2}$$

which will have a particular value for dynamically similar pump systems. Since the two pump systems are to be similar,

$$\frac{H_1}{D_1^2 N_1^2} = \frac{H_2}{D_2^2 N_2^2}$$

So, 
$$D_2^2 = \frac{H_2 D_1^2 N_1^2}{N_2^2 H_1}$$

$$= \frac{16.9 \times (11.8 \times 10^{-2})^2 (1450/60)^2}{(1157/60)^2 \times 14.7}$$

from which, $D_2 = 15.86$ cm

i.e. the new pump must have an impeller of outside diameter = <u>15.86 cm</u>, which turns at 1157 rev min$^{-1}$.

### 6.1.3 System characteristic, theoretical and effective characteristics of a centrifugal pump, and flow rate at perfect match

*Problem*

Consider a centrifugal pump located close to the base of a large tank, which is open to atmosphere, and from which water is to be pumped through 60 m of smooth pipe (inside diameter = 11.40 cm), in the course of which it will be elevated by 8.75 m, to discharge freely into the atmosphere. Calculate data for, and plot, the system characteristic (a graph of $H_{syst}$ versus $Q$) over a range of flow rates ($Q$) varying from 0.00 to 0.050 m$^3$ s$^{-1}$.

The pump has an impeller of 9.60 cm outside diameter, carries blades 1.25 cm wide which slope backwards to make angles of 35° between tangents to their outer tips and the outer circumference of the impeller. If the impeller is to rotate at 1750 rev min$^{-1}$, calculate data for, and plot on the same graph as the system characteristic, and over the same range of flow rates as above, the theoretical pump characteristic (a graph of $H_{theor}$ versus $Q$), the effective pump characteristic ($H_{eff}$ versus $Q$), and deduce the rate (in m$^3$ s$^{-1}$) at which fluid will flow in the system. Blade thickness may be neglected, and efficiency (both $\eta_{hyd}$ and overall efficiency $\eta$) versus flow rate data for the pump are given in Table 6.1. What power will be required by the driving motor?

**Table 6.1**

| $Q$ (m$^3$ s$^{-1}$) | $\eta_{hyd}$ (%) | Overall $\eta$ (%) |
|---|---|---|
| 0.00 | 65.4 | 0.0 |
| 0.01 | 71.0 | 36.1 |
| 0.02 | 71.9 | 56.0 |
| 0.03 | 67.7 | 61.0 |
| 0.04 | 57.5 | 54.1 |
| 0.05 | 39.2 | 37.0 |

*Pumping of liquids* 121

Acceleration due to gravity = 9.81 m s$^{-2}$
Density of water = 999 kg m$^{-3}$
Viscosity of water = 1.109 × 10$^{-3}$ kg m$^{-1}$ s$^{-1}$
For turbulent flow through smooth pipes, $\phi = 0.0396\, Re^{-0.25}$

*Solution*

The system head for a flow system may be defined by:

$$H_{syst} = \frac{\Delta u^2}{2\alpha g} + \frac{\Delta P}{\rho g} + \Delta z + \frac{4\phi l u^2}{dg}$$

Under the conditions in the system being considered here, $\Delta P = 0$, and $u_1 \approx 0$. So this equation reduces to:

$$H_{syst} = \Delta z + u^2 \left( \frac{1}{2g} + \frac{4\phi l}{dg} \right)$$

Now average fluid velocity,

$$u = \frac{4Q}{\pi d^2}$$

Also,

$$Re = \frac{\rho u d}{\mu}$$

$$= \frac{4\rho Q}{\pi d \mu}$$

so that

$$\phi = 0.0396 \left( \frac{4\rho Q}{\pi d \mu} \right)^{-0.25}$$

if the fluid stream is turbulent.
Thus,

$$H_{syst} = \Delta z + \frac{8}{\pi^2 d^4 g} \left[ 1 + \frac{8l}{d} \times 0.0396 \left( \frac{4\rho Q}{\pi d \mu} \right)^{-0.25} \right] Q^2$$

Substituting data,

$$H_{syst} = 8.75 + \frac{8}{\pi^2 (0.1140)^4 \times 9.81}$$

$$\times \left[ 1 + \frac{8 \times 60 \times 0.0396}{0.1140} \left( \frac{4 \times 999}{\pi \times 0.1140 \times 1.109 \times 10^{-3}} \right)^{-0.25} \times Q^{-0.25} \right] Q^2$$

$$= 8.75 + 489.2\,(1 + 2.961 Q^{-0.25}) Q^2$$

## Problems in fluid flow

This equation may be used to calculate $H_{syst}$ for any flow rate $Q$ for the system under consideration. Data obtained in this way are given in Table 6.2.

**Table 6.2**

| $Q$ (m³ s⁻¹) | 0.00 | 0.01 | 0.02 | 0.03 | 0.04 | 0.05 |
|---|---|---|---|---|---|---|
| $H_{syst}$ (m) | 8.75 | 9.26 | 10.49 | 12.32 | 14.72 | 17.63 |

The system characteristic ($H_{syst}$ versus $Q$) may now be plotted from these data (see Figure 6.1).

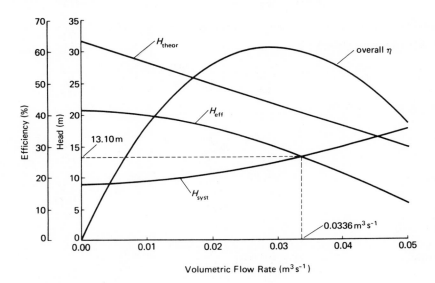

**Figure 6.1**

The theoretical head for a centrifugal pump is given by (ref. 8, 9 and 10):

$$H_{theor} = \frac{\omega^2 r_o^2}{g} - \frac{\omega Q}{2\pi g B \tan \phi'}$$

Inserting data for the system under consideration,

$$H_{theor} = \left(\frac{2\pi \times 1750}{60}\right)^2 \left(\frac{0.096^2}{9.81}\right) - \frac{2\pi \times 1750}{60 \times 2\pi \times 9.81 \times 0.0125 \times \tan 35°} Q$$

$$= 31.55 - 339.69 Q$$

This equation indicates that a plot of $H_{theor}$ versus $Q$ is a straight line. Data which obtain for the range of $Q$ values of interest are given in Table 6.3, together with corresponding values of $H_{eff} = \eta_{hyd} \cdot H_{theor}$.

*Pumping of liquids* 123

**Table 6.3**

| $Q$ (m³ s⁻¹) | $H_{theor}$ (m) | $\eta_{hyd}$ (%) | $H_{eff}$ (m) |
|---|---|---|---|
| 0.00 | 31.55 | 65.4 | 20.63 |
| 0.01 | 28.15 | 71.0 | 19.99 |
| 0.02 | 24.76 | 71.9 | 17.80 |
| 0.03 | 21.36 | 67.7 | 14.46 |
| 0.04 | 17.96 | 57.5 | 10.33 |
| 0.05 | 14.57 | 39.2 | 5.71 |

These data are plotted on the graph of Figure 6.1 as the theoretical pump characteristic ($H_{theor}$ versus $Q$) and the effective pump characteristic ($H_{eff}$ versus $Q$). It can be seen that the system characteristic and effective pump characteristic intersect where $Q = 0.0336$ m³ s⁻¹, and $H_{eff} = 13.10$ m. The maximum flow rate at which fluid will flow in this system, therefore $= 0.0336$ m³ s⁻¹. Now overall efficiency ($\eta$) may also be plotted versus flow rate (see Figure 6.1). The overall efficiency of the system when $Q$ is $0.0336$ m³ s⁻¹ $= 59.9\%$. So the power required to pump fluid at this rate,

$$= \frac{H_{eff} \rho g Q}{\eta}$$

$$= \frac{13.10 \times 999 \times 9.81 \times 0.0336}{0.599} \text{ W}$$

$$= \underline{7.20 \text{ kW}}$$

### 6.1.4 Flow rate when centrifugal pumps operate singly and in parallel

*Problem*

Two pumps develop effective heads of water at various volumetric flow rates as indicated in Table 6.4. If water has to be pumped from a large reservoir, through 98 m of smooth pipe (inside diameter $= 14.0$ cm), in the course of which it is elevated by 10.4 m, determine the flow rate at which water will move while:
(i) pump A only is in action, and
(ii) pump A and pump B are operating in parallel.
Pressure at both ends of the pipeline is approximately atmospheric.

As pumping is started from zero flow rate using pump A only, what flow rate must obtain before the delivery valve in the pipe from pump B may be opened?

Acceleration due to gravity $= 9.81$ m s⁻²
Density of water $= 999$ kg m⁻³

124  *Problems in fluid flow*

**Table 6.4**

| $Q$ (m³ s⁻¹) | $H_{eff}$ (m) for pump A | $H_{eff}$ (m) for pump B |
|---|---|---|
| 0.00 | 20.63 | 18.00 |
| 0.01 | 19.99 | 17.00 |
| 0.02 | 17.80 | 14.95 |
| 0.03 | 14.46 | 11.90 |
| 0.04 | 10.33 | 8.10 |
| 0.05 | 5.71 | 3.90 |

Viscosity of water $= 1.109 \times 10^{-3}$ kg m⁻¹ s⁻¹
For turbulent flow in smooth pipes, $\phi = 0.0396\, Re^{-0.25}$

*Solution*

The system head for a flow system may be defined by:

$$H_{syst} = \Delta z + \Delta\left(\frac{u^2}{2\alpha g}\right) + \frac{\Delta P}{\rho g} + \frac{4\phi l u^2}{dg}$$

which because $\Delta P = 0$ and $u_1 \approx 0$ for the system under consideration, reduces to,

$$H_{syst} = \Delta z + \frac{u^2}{2\alpha g} + \frac{4\phi l u^2}{dg} \tag{6.9}$$

If the fluid stream is turbulent,

$$\phi = 0.0396\, Re^{-0.25}$$

$$= 0.0396\left(\frac{4Q\rho}{\pi d \mu}\right)^{-0.25} \tag{6.10}$$

Also, $\quad \alpha = 1 \tag{6.11}$

and $\quad u = \dfrac{4Q}{\pi d^2} \tag{6.12}$

So substituting from equations (6.11) and (6.12) into equation (6.9)

$$H_{syst} = \Delta z + \frac{8}{\pi^2 d^4 g}\left(1 + \frac{8\phi l}{d}\right)Q^2$$

Then using equation (6.10),

$$H_{syst} = \Delta z + \frac{8}{\pi^2 d^4 g}\left(1 + \frac{8l}{d} 0.0396 \left(\frac{4\rho}{\pi d \mu}\right)^{-0.25} Q^{-0.25}\right)Q^2$$

Introducing data for the system under consideration,

$$H_{syst} = 10.4 + \frac{8}{\pi^2(0.14)^4 \times 9.81}$$

$$\times \left[ 1 + \frac{8 \times 98 \times 0.0396}{0.14} \left( \frac{4 \times 999}{\pi \times 0.14 \times 1.109 \times 10^{-3}} \right)^{-0.25} \times Q^{-0.25} \right] Q^2$$

$$= 10.4 + 215.08(1 + 4.145Q^{-0.25})Q^2$$

Now this equation may be used to calculate $H_{syst}$ values for the flow rates of interest. Data obtained in this way are shown in Table 6.5.

Table 6.5

| $Q$ (m³ s⁻¹) | 0.00 | 0.02 | 0.03 | 0.04 | 0.05 | 0.06 |
|---|---|---|---|---|---|---|
| $H_{syst}$ (m) | 10.40 | 11.43 | 12.52 | 13.93 | 15.65 | 17.66 |

These data, plotted on the graph of Figure 6.2, constitute the system characteristic. This curve intersects the effective pump characteristic for pump A only (obtained by plotting data given) at a volumetric flow rate = 0.0339 m³ s⁻¹. This, therefore, will be the rate at which fluid will move while pump A only is in action.

The effective pump characteristic for pumps A and B in parallel is obtained by adding the flow rates at which fluid is delivered by pumps A and B separately for a number of fluid heads. The combined characteristic, obtained

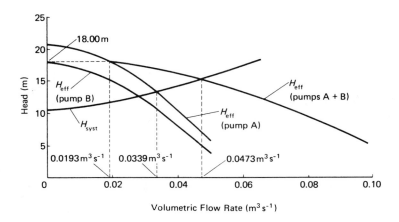

Figure 6.2

126  Problems in fluid flow

in this way, is also shown in Figure 6.2. The combined characteristic curve intersects the system characteristic at a flow rate of 0.0473 m³ s⁻¹, so this will be the flow rate at which liquid will be delivered with pumps A and B operating in parallel.

From Figure 6.2, it may be seen that the effective head which will be developed by pump B with its delivery valve closed = 18.00 m. The effective head developed by pump A will not have fallen to this value until pump A is delivering water at 0.0193 m³ s⁻¹. So to avoid useless recycling of water back to the inlet of pump A through pump B, the delivery valve for pump B must not be opened until pump A is delivering water at 0.0193 m³ s⁻¹.

To check for turbulence, we may calculate $Re$ at a low flow rate (0.010 m³ s⁻¹ for example).

$$Re = \frac{\rho u d}{\mu}$$

$$= \frac{4Q\rho}{\pi d \mu}$$

$$= \frac{4 \times 0.010 \times 999}{\pi \times 0.14 \times 1.109 \times 10^{-3}}$$

$$= 8.19 \times 10^4$$

This is greater than 4000, so flow will be turbulent, and use of equation (6.10) was appropriate.

### 6.1.5 Pumping with a reciprocating pump

*Problem*

Water (density = 1000 kg m⁻³) is to be pumped from a reservoir by a single acting reciprocating pump (cylinder diameter = 150 mm, stroke length = 220 mm). The water surface in the reservoir lies 4.60 m below the inlet port of the pump, and water is drawn into the pump through a pipe 7.8 m long and 4.5 cm in diameter. If the piston of the pump moves in simple harmonic motion, determine the maximum speed (in Hz) at which the pump may be driven without causing cavitation in the suction pipe.

Atmospheric pressure on the day = 105.4 kPa

Saturation vapour pressure of water at the temperature prevailing = 10.85 kPa

Acceleration due to gravity = 9.81 m s⁻²

*Pumping of liquids* 127

*Solution*

The tendency for cavitation to occur will be greatest
(i) at the inlet to the cylinder, because static pressure in the suction pipe will be at a minimum there, and
(ii) at the commencement of the suction stroke, because acceleration of the piston will be a maximum then, so that suction pressure exerted on fluid by pump action will also be a maximum at that time.

Let position, $x$, of the piston be measured from the centre of the cylinder where $x = 0$. Also, let the maximum frequency of oscillation of the piston

$$= N \text{ Hz},$$
$$= 2\pi N \text{ rad s}^{-1}$$

If the piston is considered to start at time $t = 0$ from the inlet end of the cylinder, at time $t = 0$,

$$x = -\frac{220 \times 10^{-3}}{2} \text{ m}$$
$$= -0.110 \text{ m}$$

So at time $t$, $\quad x(t) = -0.110\cos(2\pi Nt) \text{ m}$

Thus, velocity of the piston at time $t$,

$$\frac{dx}{dt} = 0.110 \times 2\pi N \sin(2\pi Nt) \text{ m s}^{-1}$$

and acceleration of the piston then,

$$\frac{d^2x}{dt^2} = 0.110(2\pi N)^2 \cos(2\pi Nt) \text{ m s}^{-2}$$
$$= 4.343 N^2 \cos(2\pi Nt) \text{ m s}^{-2}$$

Acceleration will be a maximum when

$$\cos(2\pi Nt) = 1$$

i.e. when $\quad t = 0$

Thus, maximum acceleration of the piston $= 4.343 N^2 \text{ m s}^{-2}$
This will cause acceleration of fluid in the suction pipe to be a maximum, and this will

$$= \frac{\frac{1}{4}\pi(150 \times 10^{-3})^2}{\frac{1}{4}\pi(4.5 \times 10^{-2})^2} \times 4.343 N^2 \text{ m s}^{-2}$$
$$= 48.26 N^2 \text{ m s}^{-2}$$

Now the corresponding suction pressure is required. To this end, mass of fluid

## 128  Problems in fluid flow

in the suction pipe

$$= \frac{\pi(4.5 \times 10^{-2})^2 \times 7.8 \times 1000}{4} \text{ kg}$$

$$= 12.41 \text{ kg}$$

So the force exerted at maximum acceleration

$$= 12.41 \times 48.26 \, N^2 \text{ N}$$

$$= 598.9 \, N^2 \text{ N}$$

so that maximum suction pressure

$$= \frac{4 \times 598.9 \, N^2}{\pi(4.5 \times 10^{-2})^2} \text{ Pa}$$

$$= 3.766 \times 10^5 \, N^2 \text{ Pa}$$

Now minimum pressure in the suction pipe will occur at the entry to the cylinder, and

$$= (105.4 \times 10^3) - (4.60 \times 1000 \times 9.81) - (3.766 \times 10^5 N^2) \text{ Pa}$$

$$= 6.027 \times 10^4 - 3.766 \times 10^5 \, N^2 \text{ Pa}$$

Cavitation will occur when this falls to the saturation vapour pressure of the water, i.e. when:

$$6.027 \times 10^4 - 3.766 \times 10^5 \, N^2 = 10.85 \times 10^3$$

from which, $\quad N = 0.362 \text{ Hz}$

i.e. maximum frequency of oscillation of the piston if cavitation is to be avoided = <u>0.362 Hz</u>

### 6.1.6  Pumping with an air-lift pump

*Problem*

Sulphuric acid is to be transferred by an air-lift pump from a large holding tank to discharge into a tanker. Air will be admitted to the rising main 4.30 m below the discharge point and 1.65 m below the surface of acid in the holding tank. If energy of the compressed air can be used in the process with 93% efficiency, calculate the mass of air which will be required to fill the tanker from empty with sulphuric acid if its tank is cylindrical (diameter = 2.45 m, length = 10.5 m). Calculate also the mass of air required per kilogram of acid transferred.

Density of sulphuric acid = $1830 \text{ kg m}^{-3}$

Atmospheric pressure on the day = 104.2 kPa
Temperature on the day = 11 °C
Molar mass of air = $28.8 \times 10^{-3}$ kg mol$^{-1}$
Universal gas constant = 8.314 J mol$^{-1}$ K$^{-1}$
Acceleration due to gravity = 9.81 m s$^{-2}$

*Solution*

The mass of air required to raise unit mass of liquid in an air-lift pump, (cf. ref. 11),

$$\frac{m_{air}}{m_{liq}} = \frac{g h_r}{\eta P_{atm} v_{atm} \ln[h_{atm}/(h_{atm} + h_s)]} \qquad (6.13)$$

(Note: This equation gives a negative value to the ratio $m_{air}/m_{liq}$ because work is done by the air as it expands.)
If the air behaves ideally, $\qquad PV = nRT$

so that, $\qquad\qquad\qquad \dfrac{RT}{M_{air}} = P \dfrac{V}{nM_{air}}$

$\qquad\qquad\qquad\qquad\qquad = Pv$

$\qquad\qquad\qquad\qquad\qquad =$ constant

$\qquad\qquad\qquad\qquad\qquad = P_{atm} v_{atm}$

Using this equality to substitute for the product $P_{atm} v_{atm}$ in equation (6.13),

$$m_{air} = \frac{gh_r M_{air} m_{liq}}{\eta RT \ln[h_{atm}/(h_{atm} + h_s)]} \qquad (6.14)$$

Now, the mass of liquid to be transferred,

$$m_{liq} = \frac{\pi \times 2.45^2 \times 10.5 \times 1830}{4} \text{ kg}$$

$$= 9.059 \times 10^4 \text{ kg}$$

Also because, $\qquad P_{atm} = h_{atm} \rho_{liq} g$

$$h_{atm} = \frac{104.2 \times 10^3}{1830 \times 9.81} = 5.804 \text{ m}$$

Since $\qquad\qquad h_s = 1.65$ m

$\qquad\qquad\qquad h_r = 4.30 - 1.65$ m

$\qquad\qquad\qquad\quad = 2.65$ m

So substituting calculated and given data into equation (6.14), the mass of air required for the transfer operation,

$$m_{air} = \frac{9.81 \times 2.65 \times 28.8 \times 10^{-3} \times 9.059 \times 10^4}{0.93 \times 8.314 \times 284 \times \ln[5.804/(5.804 + 1.65)]} \text{ kg}$$

$$= \underline{-123.4 \text{ kg}}$$

(The negative sign is obtained because work is done by the air as it expands, so it is the magnitude of this quantity which is significant here.)

The mass of air required per kilogram of acid transferred,

$$= \frac{123.4}{9.059 \times 10^4} \text{ kg}$$

$$= \underline{0.001\,36 \text{ kg}}$$

## 6.2 Student Exercises

1  A centrifugal pump, with a minimum net positive suction head = 1.20 m of water, is to be used to pump pure nitric acid from a holding tank, in which pressure at the free surface of liquid is atmospheric, into carboys. Acid is drawn through a resistant, high-silicon iron suction pipe (inside diameter = 2.80 cm, length = 7.15 m). The Stanton–Pannell friction factor, $\phi$, may be considered = 0.0056 under all conditions to be expected. If it is possible to keep the pump primed, calculate:

(i) the maximum height above the liquid surface in the holding tank to which acid may be drawn in the suction pipe,

(ii) the maximum rate (in $dm^3\,s^{-1}$) at which acid may be pumped without risk of cavitation if the pump is located 4.32 m above liquid in the holding tank, and

(iii) the minimum diameter of suction pipe required if acid is to be pumped at 2.50 $dm^3\,s^{-1}$ without risk of cavitation and with the pump located 4.32 m above liquid in the holding tank. Neglect changes in kinetic energy of the acid streams.

Atmospheric pressure on the day = 103.7 kPa
Density of water = 999 $kg\,m^{-3}$
Density of pure nitric acid = 1531 $kg\,m^{-3}$
Saturation vapour pressure of pure nitric acid = 3.55 kPa
Acceleration due to gravity = 9.81 $m\,s^{-2}$

2  A centrifugal pump develops an 11.75 m head of water (density = 999 $kg\,m^{-3}$) while pumping the liquid at 40.0 $dm^3\,s^{-1}$ with a rotor turning rate = 850 rev $min^{-1}$. Calculate:

(i) the specific speed of the pump, and
(ii) the power being imparted to the liquid.

If the pump, operating at the same power, is now to be used to pump oil (density = 886 kg m$^{-3}$) under dynamically similar conditions, calculate:
(iii) the head of oil which the pump will develop, and
(iv) the volumetric flow rate of oil which will obtain.

Acceleration due to gravity = 9.81 m s$^{-2}$

3   Water is conducted from a canal, through 38.7 m of smooth pipe (inside diameter = 16.5 cm), to a centrifugal pump situated 9.45 m below the free surface of water in the canal. Water then passes through 286 m of smooth pipe (inside diameter = 12.0 cm) to discharge freely at a point 47.6 m above the pump. Calculate data for, and plot, the system characteristic (a graph of $H_{syst}$ versus $Q$) over a range of flow rates, $Q$ from 0.000 to 0.040 m$^3$ s$^{-1}$.

The pump has an impeller of 15.0 cm outside diameter which carries blades 2.25 cm wide, and which slope backwards to make angles of 35° between tangents to their outer tips and the outer circumference of the impeller. If the impeller is to rotate at 1450 rev min$^{-1}$, calculate values of theoretical head for the pump also at volumetric flow rates from 0.000 to 0.040 m$^3$ s$^{-1}$.

The hydraulic efficiency of the pump at various flow rates is given in Table 6.6. Use these to calculate data for the effective pump characteristic ($H_{eff}$ versus $Q$). Then, on the same diagram as the system characteristic, plot the effective pump characteristic and deduce the maximum rate (in dm$^3$ s$^{-1}$) at which water will be able to flow in the system. Thickness of the impeller blades may be neglected.

**Table 6.6**

| $Q$ (m$^3$ s$^{-1}$) | 0.000 | 0.010 | 0.020 | 0.030 | 0.040 |
|---|---|---|---|---|---|
| $\eta_{hyd}$ (%) | 82.5 | 83.9 | 83.0 | 78.3 | 68.8 |

Density of water = 999 kg m$^{-3}$
Viscosity of water = 1.22 × 10$^{-3}$ kg m$^{-1}$ s$^{-1}$
Acceleration due to gravity = 9.81 m s$^{-2}$
For turbulent flow through smooth pipes, $\phi = 0.0396\ Re^{-0.25}$

4   Water must be pumped out of a mine at flow rates up to 0.035 m$^3$ s$^{-1}$. Water will be raised through a vertical, smooth pipeline (length = 1375 m, inside diameter = 18.0 cm) by a multi-stage series of centrifugal pumps situated at the bottom of the mine. Each stage of the pump system it is intended to use develops effective heads ($H_{eff}$) of water at various volumetric flow rates $Q$ as indicated in Table 6.7. If the difference in pressure which obtains between the two ends of the pipe system is neglected, find the number of pumping stages it will be necessary to incorporate in the pump system.

## Table 6.7

| $Q$ (m³ s⁻¹) | 0.00 | 0.01 | 0.02 | 0.03 | 0.04 | 0.05 |
|---|---|---|---|---|---|---|
| $H_{eff}$ (m) | 44.8 | 43.7 | 41.0 | 38.8 | 34.5 | 28.9 |

Density of water = 998 kg m⁻³
Viscosity of water = $1.027 \times 10^{-3}$ kg m⁻¹ s⁻¹
Acceleration due to gravity = 9.81 m s⁻²
For turbulent flow through smooth pipes, $\phi = 0.0396\, Re^{-0.25}$

5  It is necessary to pump water at 40.0 dm³ s⁻¹ for sluicing purposes. Water will be drawn from a large tank, and be pumped by a battery of identical, low capacity, centrifugal pumps arranged to pump in parallel. The pipeline will be smooth, 114 m long and of inside diameter = 8.50 cm. Water will be elevated 5.18 m during the operation, and pressure at both ends of the pipeline will be approximately atmospheric. If the effective head ($H_{eff}$) developed by a single pump varies with flow rate as shown by data in Table 6.8, find the minimum number of pumps it will be necessary to install in the battery.

## Table 6.8

| $Q$ (dm³ s⁻¹) | 0.0 | 2.0 | 4.0 | 6.0 | 8.0 | 10.0 |
|---|---|---|---|---|---|---|
| $H_{eff}$ (m) | 57.1 | 54.9 | 51.3 | 46.1 | 38.5 | 27.6 |

Density of water = 999 kg m⁻³
Viscosity of water = $1.12 \times 10^{-3}$ kg m⁻¹ s⁻¹
Acceleration due to gravity = 9.81 m s⁻²
For turbulent flow through smooth pipes, $\phi = 0.0396\, Re^{-0.25}$

6  A single-acting, reciprocating pump (cylinder diameter = 60 mm, stroke length = 100 mm) is to be used to pump 2-propanol (density = 785.5 kg m⁻³) out of large drums. The suction pipe will be 5.20 m long and of inside diameter = 1.5 cm. If the piston of the pump moves in simple harmonic motion at a frequency of 0.70 Hz,
(i) what will be the maximum distance to which the liquid surface in the drums will be able to fall below pump level before cavitation will occur in the suction pipe, and
(ii) what will be the average volumetric flow rate at which propanol will flow in the system?

Atmospheric pressure on the day = 102.7 kPa

*Pumping of liquids* 133

Saturation vapour pressure of 2-propanol at the temperature prevailing
= 3.07 kPa

Acceleration due to gravity = 9.81 m s$^{-2}$

7   Water is sometimes pumped from wells by means of an air-lift pump so as to oxidise iron salts in solution to ferric hydroxide prior to sedimentation of the hydroxide and use of the water.

It will be acceptable to use 10 g of air per kilogram of water in order to pump water from a well with an air-lift pump. If water must be lifted 15.6 m above the level of water in the well, calculate how far below the level of water in the well it will be necessary to feed air into the rising main. The energy of the compressed air can be used with 94% efficiency.

Density of water = 1000 kg m$^{-3}$

Atmospheric pressure on the day = 100.1 kPa

Temperature of the system = 8 °C

Molar mass of air = 28.8 × 10$^{-3}$ kg mol$^{-1}$

Universal gas constant = 8.314 J mol$^{-1}$ K$^{-1}$

Acceleration due to gravity = 9.81 m s$^{-2}$

## 6.3  References

References are abbreviated as follows:

C & R (1) = J. M. Coulson & J. F. Richardson, Chemical Engineering, Vol. 1, 3rd edition, Pergamon, 1977

McC, S & H = W. L. McCabe, J. C. Smith and P. Harriott, Unit Operations of Chemical Engineering, 4th edition, McGraw-Hill, 1985

D, G & S = J. F. Douglas, J. M. Gasiorek & J. A. Swaffield, Fluid Mechanics, Pitman, 1980

1. C & R (1), page 139
2. McC, S & H, page 171
3. D, G & S, page 650
4. C & R (1), page 43
5. D, G & S, page 254
6. C & R (1), page 136
7. D, G & S, page 601
8. C & R (1), page 135
9. McC, S & H, page 179
10. D, G & S, page 562
11. C & R (1), page 146

## 6.4  Notation

| Symbol | Description | Unit |
|---|---|---|
| $B$ | width of the blades of a pump's impeller | m |
| $d$ | diameter of a pipe | m |
| $D$ | diameter of the impeller of a centrifugal pump | m |
| $D_1$ | diameter of the impeller of pump no. 1 | m |

134  *Problems in fluid flow*

| Symbol | Description | Unit |
|---|---|---|
| $D_2$ | diameter of the impeller of pump no. 2 | m |
| $g$ | acceleration due to gravity | $m\,s^{-2}$ |
| $h_{atm}$ | height of a column of the liquid being pumped which could be supported by atmospheric pressure | m |
| $h_f$ | head loss due to friction | m |
| $h_r$ | net vertical distance through which liquid is raised in an air-lift pump system | m |
| $h_s$ | vertical distance between free liquid surface in the feed tank and the point at which air enters the rising main in an air-lift pump system | m |
| $H$ | head of fluid developed by a pump | m |
| $H_1$ | head of fluid developed by pump no. 1 | m |
| $H_2$ | head of fluid developed by pump no. 2 | m |
| $H_{eff}$ | actual (effective) head of fluid developed by a centrifugal pump | m |
| $H_{syst}$ | head of fluid required to drive fluid through a system | m |
| $H_{theor}$ | theoretical head of fluid which could be developed by a centrifugal pump | m |
| $l$ | length of a pipe | m |
| $m_{air}$ | mass of air used to raise liquid in an air-lift pump | kg |
| $m_{liq}$ | mass of liquid raised in an air-lift pump | kg |
| $M_{air}$ | molar mass of air | $kg\,mol^{-1}$ |
| $n$ | number of moles of a gas | |
| $N$ | pump speed | $s^{-1}$ |
| $N_1$ | speed of pump no. 1 | $s^{-1}$ |
| $N_2$ | speed of pump no. 2 | $s^{-1}$ |
| $N_s$ | specific speed of a centrifugal pump | $m^{0.75}\,s^{-1.5}$ |
| N.P.S.H. | net positive suction head | m |
| $P$ | pressure of a gas | Pa |
| $P_1$ | pressure in a fluid at station 1 | Pa |
| $P_{atm}$ | atmospheric pressure | Pa |
| $P_{liq}$ | saturation vapour pressure of a liquid | Pa |
| $Q$ | volumetric flow rate of liquid | $m^3\,s^{-1}$ |
| $r_o$ | radius of the outer circumference of a pump's impeller | m |
| $R$ | the Universal Gas Constant | $J\,mol^{-1}\,K^{-1}$ |
| $Re$ | Reynolds number for fluid flow in a pipe | |
| $t$ | time | s |
| $T$ | temperature of a body of gas | K |
| $u$ | average velocity of fluid in a pipe | $m\,s^{-1}$ |
| $u_1$ | average velocity of fluid at station 1 | $m\,s^{-1}$ |
| $u_2$ | average velocity of fluid at station 2 | $m\,s^{-1}$ |
| $v$ | specific volume of a gas at pressure P | $m^3\,kg^{-1}$ |

| Symbol | Description | Unit |
|---|---|---|
| $v_{atm}$ | specific volume of air at atmospheric pressure | $m^3\,kg^{-1}$ |
| $V$ | volume occupied by $n$ moles of gas at pressure $P$ | $m^3$ |
| $x$ | position of the piston in the cylinder of a reciprocating pump | m |
| $z_i$ | fluid head on the inlet side of a pump | m |
| $\alpha$ | kinetic energy correction factor | |
| $\Delta P$ | pressure difference across a system | Pa |
| $\Delta u^2$ | $u_2^2 - u_1^2$ | $m^2\,s^{-2}$ |
| $\Delta z$ | the vertical lift in a pumped system | m |
| $\eta$ | overall efficiency of a pump | |
| $\eta_{hyd}$ | hydraulic efficiency of a centrifugal pump | |
| $\mu$ | viscosity of a fluid | $kg\,m^{-1}\,s^{-1}$ |
| $\rho$ | density of a fluid | $kg\,m^{-3}$ |
| $\rho_{liq}$ | density of the liquid being pumped | $kg\,m^{-3}$ |
| $\phi$ | Stanton–Pannell friction factor | |
| $\phi'$ | angle between the outer tips of the blades of a pump's impeller and a tangent to the outer circumference of the impeller at the same point, the angle being measured backwards with respect to the direction of rotation. | |
| $\omega$ | angular velocity of a pump's impeller | $rad\,s^{-1}$ |

# 7 Flow through packed beds

## 7.1 Worked Examples

### 7.1.1 Determination of particle size and specific surface area for a sample of powder

*Problem*

The density of a sample of cement is found to be 3100 kg m$^{-3}$. 25.4 g of this cement is packed into a permeability cell (diameter = 2.50 cm) to form a bed of powder 2.73 cm thick. When a pressure drop of 12.5 cm of mercury is established across the bed, it is found that 12 min 54 s are required to drive 250 cm$^3$ of air through the bed. Calculate the average diameter of particles in the cement (assuming that they are spherical), and the area of surface per gram of cement.

Viscosity of air = $1.83 \times 10^{-5}$ kg m$^{-1}$ s$^{-1}$
Density of mercury = $1.355 \times 10^4$ kg m$^{-3}$
Kozeny's constant = 5.0
Acceleration due to gravity = 9.81 m s$^{-2}$

*Solution*

The volume occupied by solid in the sample of cement

$$= \frac{25.4 \times 10^{-3}}{3100} \text{ m}^3 = 8.194 \text{ cm}^3$$

The volume of the particle bed formed by this material

$$= \frac{\pi (2.50 \times 10^{-2})^2 \times 2.73 \times 10^{-2}}{4} \text{ m}^3$$

$$= 13.401 \text{ cm}^3$$

Thus, the fractional voidage of the bed,

$$e = \frac{13.401 - 8.194}{13.401} = 0.3886$$

The volume flow rate of air through the bed

$$= \frac{250}{774} = 0.3230 \text{ cm}^3 \text{ s}^{-1}$$

The cross-sectional area of the permeability cell

$$= \frac{\pi(2.50 \times 10^{-2})^2}{4} \text{ m}^2$$

$$= 4.909 \text{ cm}^2$$

Thus, the average superficial linear flow rate of air through the bed,

$$u = \frac{0.3230}{4.909} = 6.580 \times 10^{-2} \text{ cm s}^{-1}$$

The pressure drop across the bed,

$$-\Delta P = 12.5 \times 10^{-2} \times 1.355 \times 10^4 \times 9.81 \text{ Pa}$$

$$= 16.62 \text{ kPa}$$

Using the Carman–Kozeny equation for flow through a packed bed of particles (ref. 1 and 2):

$$u = \frac{1}{K''} \cdot \frac{e^3}{S^2(1-e)^2} \cdot \frac{(-\Delta P)}{\mu l}$$

$$S^2 = \frac{0.3886^3 \times 16.62 \times 10^3}{5.0 \times 6.580 \times 10^{-2}(1-0.3886)^2 \times 1.83 \times 10^{-5} \times 2.73 \times 10^{-2}} \text{ m}^{-2}$$

so,

$$S = 1.260 \times 10^5 \text{ m}^{-1}$$

Now,

$$S = \frac{4\pi r^2}{\frac{4}{3}\pi r^3}$$

$$= 6/d$$

So, the average particle diameter,

$$d = \frac{6}{1.260 \times 10^5} \text{ m}$$

$$= \underline{47.62 \ \mu m}$$

The surface area of such a particle

$$= \pi d^2$$

$$= \pi(47.62 \times 10^{-6})^2 \text{ m}^2$$

$$= 7.124 \times 10^{-9} \text{ m}^2$$

138  *Problems in fluid flow*

The mass of such a particle

$$= \tfrac{4}{3}\pi r^3 \rho_s$$
$$= \frac{4\pi}{3}\left(\frac{47.62 \times 10^{-6}}{2}\right)^3 \times 3100 \text{ kg}$$
$$= 1.753 \times 10^{-10} \text{ kg}$$

So the surface area per gram of cement

$$= \frac{7.124 \times 10^{-9} \times 10^{-3}}{1.753 \times 10^{-10}} \text{ m}^2$$
$$= \underline{406.4 \text{ cm}^2}$$

## 7.1.2 Rate of flow of fluid through a packed bed

*Problem*

A column for gas–liquid contacting is packed with a random array of Raschig rings (length = 6.35 mm, outside diameter = 6.35 mm, wall thickness = 0.76 mm). Rings are packed such that each cubic metre of column contains $3.023 \times 10^6$ rings. The column is 2.65 m in diameter, and the bed of rings in it is 24.5 m high. If the bed is prone to flooding with solvent hydrocarbon, calculate the initial volumetric flow rate at which hydrocarbon may be drained from the column if gas flow is stopped when the bed has become full of liquid, and an additional layer of liquid 0.65 m deep lies on top of the bed.

Viscosity of hydrocarbon solvent = $2.5 \times 10^{-3}$ kg m$^{-1}$ s$^{-1}$
Density of hydrocarbon solvent = 897 kg m$^{-3}$
Acceleration due to gravity = 9.81 m s$^{-2}$
Kozeny's constant = 5.1

*Solution*

Volume of solid material in each Raschig ring

$$= \frac{\pi(6.35 \times 10^{-3})^2 \times 6.35 \times 10^{-3}}{4}$$
$$- \frac{\pi[(6.35 - 2 \times 0.76) \times 10^{-3}]^2 \times 6.35 \times 10^{-3}}{4} \text{ m}^3$$
$$= 8.475 \times 10^{-8} \text{ m}^3$$

So fractional voidage of the bed,
$$e = \frac{1 - (8.475 \times 10^{-8} \times 3.023 \times 10^6)}{1}$$
$$= 0.7438$$

Surface area of each Raschig ring
$$= \pi(6.35 \times 10^{-3})(6.35 \times 10^{-3})$$
$$+ \pi(6.35 - 2 \times 0.76) \times 10^{-3} \times 6.35 \times 10^{-3}$$
$$+ \frac{2\pi(6.35 \times 10^{-3})^2}{4}$$
$$- \frac{2\pi[(6.35 - 2 \times 0.76) \times 10^{-3}]^2}{4} \, m^2$$
$$= 2.497 \times 10^{-4} \, m^2$$

So,
$$S = \frac{2.497 \times 10^{-4}}{8.475 \times 10^{-8}} = 2946 \, m^{-1}$$

Now the total height of fluid in the column when it is flooded
$$= 24.5 + 0.65 = 25.15 \, m$$

So pressure drop across the bed,
$$-\Delta P = 25.15 \times 897 \times 9.81 \, Pa = 221.3 \, kPa$$

Now, using the Carman–Kozeny equation (ref. 1 and 2), initial average fluid velocity as fluid drains out of the bed
$$u = \frac{1}{K''} \cdot \frac{e^3}{S^2(1-e)^2} \cdot \frac{(-\Delta P)}{\mu l}$$

$$= \frac{0.7438^3 \times 221.3 \times 10^3}{5.1 \times 2946^2 (1 - 0.7438)^2 \times 2.5 \times 10^{-3} \times 24.5} \, m\,s^{-1}$$

$$= 0.5117 \, m\,s^{-1}$$

So, initial volumetric flow rate of solvent out of the column
$$= \frac{0.5117 \times \pi \times 2.65^2}{4} \, m^3\,s^{-1}$$
$$= \underline{2.822 \, m^3\,s^{-1}}$$

## 7.1.3 Determination of the pressure drop to drive fluid through a packed bed of Raschig rings, then of similar sized spheres, and the determination of total area of surface presented with the two types of packing

*Problem*

(i) A reactor column (diameter = 2.0 m) is to have oxygen flowing in it counter current to a falling stream of liquid. If the column is packed with a random array of porcelain Raschig rings (length = 50.8 mm, outside diameter = 50.8 mm, wall thickness = 6.35 mm, and with 5790 of them packed in each cubic metre of bed) to form a bed 18 m high, calculate the pressure drop (in cm of water) which will obtain over the bed when oxygen is driven through the column at $10^4$ m³ h⁻¹ before liquid is admitted to the column. What area of surface is presented for contact between gas and liquid in the column?

(ii) If the column were packed with a random array of spheres (diameter = 50.8 mm) to give a fractional voidage of 0.432, calculate the pressure drop (in cm of water) which would be required to maintain oxygen flow at a rate of $10^4$ m³ h⁻¹ while the column is dry, and the area of surface which would then be available for contacting fluids.

Viscosity of oxygen = $2.02 \times 10^{-5}$ kg m⁻¹ s⁻¹
Density of water = 998 kg m⁻³
Kozeny's constant = 5.1
Acceleration due to gravity = 9.81 m s⁻²

*Solution*

(i) Cross-sectional area of the reactor column

$$= \pi \times 2.0^2/4 = 3.142 \text{ m}^2$$

so the average superficial fluid velocity through the bed

$$u = \frac{10^4}{3.142 \times 3600} = 0.8842 \text{ m s}^{-1}$$

Now the volume of porcelain in each Raschig ring

$$= \frac{\pi (50.8 \times 10^{-3})^2 \times 50.8 \times 10^{-3}}{4}$$

$$- \frac{\pi [(50.8 - 2 \times 6.35) \times 10^{-3}]^2 \times 50.8 \times 10^{-3}}{4} \text{ m}^3$$

$$= 4.505 \times 10^{-5} \text{ m}^3$$

So, the fractional voidage of the bed,

$$e = \frac{1 - (4.505 \times 10^{-5} \times 5790)}{1} = 0.7392$$

The surface area of each Raschig ring

$$= \pi(50.8 \times 10^{-3})^2$$
$$+ \pi[(50.8 - 2 \times 6.35) \times 10^{-3}] \times 50.8 \times 10^{-3}$$
$$+ \frac{2\pi(50.8 \times 10^{-3})^2}{4}$$
$$- \frac{2\pi[(50.8 - 2 \times 6.35) \times 10^{-3}]^2}{4} \text{ m}^2$$
$$= 1.596 \times 10^{-2} \text{ m}^2$$

So, $\quad S = \dfrac{1.596 \times 10^{-2}}{4.505 \times 10^{-5}} \text{ m}^{-1}$

$$= 354.3 \text{ m}^{-1}$$

Now the Carman–Kozeny equation for flow through packed beds is (ref. 1 and 2):

$$u = \frac{1}{K''} \cdot \frac{e^3}{S^2(1-e)^2} \cdot \frac{(-\Delta P)}{\mu l}$$

so the pressure drop required to drive oxygen through the bed when dry,

$$-\Delta P = \frac{K''S^2(1-e)^2 \, \mu l u}{e^3}$$
$$= \frac{5.1 \times 354.3^2 (1 - 0.7392)^2 \times 2.02 \times 10^{-5} \times 18 \times 0.8842}{0.7392^3} \text{ Pa}$$
$$= 34.66 \text{ Pa}$$

Now, $\quad -\Delta P = \Delta h \rho_w g$

so $\quad \Delta h = \dfrac{34.66}{998 \times 9.81} \text{ m}$

$$= 0.354 \text{ cm}$$

i.e. the pressure drop which will obtain over the bed of Raschig rings
= 0.354 cm $H_2O$

## 142 Problems in fluid flow

The area of surface presented for contact between gas and liquid in the column of rings

$$= \left[\frac{1.596 \times 10^{-2} \times 5790 \times \pi \times 2.0^2 \times 18}{4}\right] + (\pi \times 2.0 \times 18)\ \text{m}^2$$

$$= \underline{5339\ \text{m}^2}$$

(ii) If the column were packed with spheres,

$$S = \frac{4\pi r^2}{\tfrac{4}{3}\pi r^3}$$

$$= \frac{6}{d}$$

$$= \frac{6}{50.8 \times 10^{-3}} = 118.1\ \text{m}^{-1}$$

In this case, pressure drop required to drive oxygen through the bed of spheres when dry would be,

$$-\Delta P = \frac{5.1 \times 118.1^2 (1-0.432)^2 \times 2.02 \times 10^{-5} \times 18 \times 0.8842}{0.432^3}\ \text{Pa}$$

$$= 91.51\ \text{Pa}$$

Since $\quad -\Delta P = \Delta h \rho_w g$

$$\Delta h = \frac{91.51}{998 \times 9.81}\ \text{m}$$

$$= 0.935\ \text{cm}$$

i.e. the pressure drop which would be required to drive oxygen through the bed of spheres = $\underline{0.935\ \text{cm}\ \mathbf{H_2O}}$

Now the volume of solid which would be packed into the bed if the particles were spheres

$$= \frac{\pi \times 2.0^2 \times 18(1-0.432)}{4}\ \text{m}^3$$

$$= 32.12\ \text{m}^3$$

So since $\quad S = 118.1\ \text{m}^{-1}$

the surface area of spheres in the bed

$$= 32.12 \times 118.1\ \text{m}^2$$

$$= 3793\ \text{m}^2$$

Including also the surface area of the tower walls, the area of surface which would be available for contacting fluids

$$= 3793 + (\pi \times 2.0 \times 18) \, m^2$$
$$= \underline{3906 \, m^2}$$

## 7.2 Student Exercises

1 A standard sample of pigment powder (density = 1266 kg m$^{-3}$) is known to consist of particles (assumed spherical) with an average diameter of 54.6 μm. If 8.50 g of the powder can be packed consistently to give a bed depth of 1.80 cm in a permeability cell which is 3.00 cm in diameter, find the time it will take to drive 2.00 dm$^3$ of air through the bed under a pressure head of 25.0 cm of oil (density = 884 kg m$^{-3}$).

Viscosity of air = $1.83 \times 10^{-5}$ kg m$^{-1}$ s$^{-1}$

Kozeny's constant = 4.9

Acceleration due to gravity = 9.81 m s$^{-2}$

2 A pipe (inside diameter = 12.0 cm) is weighed empty, then again after steel balls (density = 7996 kg m$^{-3}$) have been dumped into it to form a randomly packed bed 1.16 m deep. The increase in weight caused by addition of the balls = 63.68 kg. The bed is then flooded with water such that the overall height of the column of water through the bed = 1.48 m. If it is now found that water can percolate through the bed at 64 cm$^3$ s$^{-1}$, determine the diameter of the steel balls.

Viscosity of water = $1.25 \times 10^{-3}$ kg m$^{-1}$ s$^{-1}$

Density of water = 998 kg m$^{-3}$

Kozeny's constant = 5.0

Acceleration due to gravity = 9.81 m s$^{-2}$

3 A filter cake, which is known to consist of particles 88 μm in diameter, and have a dry density of 731 kg m$^{-3}$, is being washed with water in a plate-and-frame filter press. 15 frames, each 60 cm wide and 120 cm tall, are in use. If water is being pumped through the press at 438 dm$^3$ s$^{-1}$ under a pressure drop of 86 kPa, calculate the thickness of the filter cake if resistance of the filter medium may be neglected.

Density of particles = 1342 kg m$^{-3}$

Viscosity of water = $1.24 \times 10^{-3}$ kg m$^{-1}$ s$^{-1}$

Kozeny's constant = 5.0

## 7.3 References

1. J. M. Coulson and J. F. Richardson, Chemical Engineering, Vol. 2, 3rd edition, Pergamon, 1980, page 129
2. W. L. McCabe, J. C. Smith and P. Harriott, Unit Operations of Chemical Engineering, 4th edition, McGraw-Hill, 1985, page 137

## 7.4 Notation

| Symbol | Description | Unit |
|---|---|---|
| $d$ | diameter of a particle | m |
| $e$ | fractional voidage of a particle bed | — |
| $g$ | acceleration due to gravity | $m\,s^{-2}$ |
| $K''$ | Kozeny's constant | — |
| $l$ | thickness of a particle bed | m |
| $r$ | radius of a particle | m |
| $S$ | surface area of a particle per unit volume of the particle | $m^{-1}$ |
| $u$ | average superficial velocity of fluid through a particle bed | $m\,s^{-1}$ |
| $\Delta h$ | difference in the heights of liquid levels in a manometer | m |
| $\Delta P$ | pressure difference across a particle bed | Pa |
| $\mu$ | viscosity of a fluid | $kg\,m^{-1}\,s^{-1}$ |
| $\rho_s$ | density of the solid in a particle bed | $kg\,m^{-3}$ |
| $\rho_w$ | density of water | $kg\,m^{-3}$ |

# 8 Filtration

*Note*: The various authors of theoretical texts have developed filtration theory in diverse ways, making it difficult to quote references directly to equations developed in that literature. For this reason, a single text has been selected here for reference to basic equations, then these basic equations have been developed further below.

## 8.1 Worked Examples

### 8.1.1 Constant rate filtration in a plate-and-frame filter press when an incompressible filter cake forms. Evaluation of system parameters

*Problem*

An aqueous slurry of china clay, which forms an incompressible filter cake, and which has a mass fraction of clay = 0.0278, was filtered in a plate-and-frame filter press. 17 frames of the press were in operation, each 125 cm tall and 70 cm wide. Filtration was carried out at a constant rate of 78 dm³ s⁻¹. After 2 minutes of filtration, the pressure drop across the filter was 3.23 kPa, and after 5 minutes, it was 6.53 kPa. Calculate:
(i) the pressure drop which will obtain across the filter after 30 minutes of filtration,
(ii) the thickness of the filter cake which will have formed after 30 minutes,
(iii) the thickness of cake which is equal in resistance to the resistance of the filter medium, and
(iv) the average diameter of particles in the slurry if they are considered to be spherical.

Density of water = 998 kg m⁻³
Density of the china clay = 2230 kg m⁻³
Density of dry filter cake = 1324 kg m⁻³
Viscosity of water = 1.002 × 10⁻³ kg m⁻¹ s⁻¹

*Solution*

(i) For filtration through cloth and cake combined, by manipulating the

Carman–Kozeny equation, it may be shown that:
$$\frac{dV}{dt} = \frac{A^2(-\Delta P)}{vr\mu(V + LA/v)} \quad \text{(ref. 1)}$$

From this, for constant-rate filtration, when
$$\frac{dV}{dt} = \frac{V}{t}$$

so that
$$V = \frac{dV}{dt} \cdot t$$

it follows that:
$$-\Delta P = \frac{r\mu v}{A^2}\left(\frac{dV}{dt}\right)^2 t + \frac{r\mu L}{A}\left(\frac{dV}{dt}\right) \quad (8.1)$$

$$\equiv K_1\left(\frac{dV}{dt}\right)^2 t + K_2\left(\frac{dV}{dt}\right) \quad (8.2)$$

Now inserting data, after 2 minutes of filtration,
$$3.23 \times 10^3 = K_1 \times (78 \times 10^{-3})^2 \times 120 + K_2 \times 78 \times 10^{-3} \quad (8.3)$$

After 5 minutes,
$$6.53 \times 10^3 = K_1 \times (78 \times 10^{-3})^2 \times 300 + K_2 \times 78 \times 10^{-3} \quad (8.4)$$

By solving simultaneous equations (8.3) and (8.4), it may be shown that
$$K_1 = 3.013 \times 10^3 \text{ kg m}^{-7}\text{s}^{-1}$$
and
$$K_2 = 1.321 \times 10^4 \text{ kg m}^{-4}\text{s}^{-1}$$

So, using equation (8.2) again, after 30 minutes of operation, pressure drop,
$$-\Delta P = 3.013 \times 10^3 (78 \times 10^{-3})^2 \times 30 \times 60 + 1.321 \times 10^4 \times 78 \times 10^{-3} \text{ Pa}$$
$$= \underline{34.03 \text{ kPa}}$$

(ii) The thickness of filter cake (ref. 2)
$$l = \frac{v \cdot V}{A} \quad (8.5)$$

Now the volume of filtrate which will have collected after 30 minutes,
$$V = 78 \times 10^{-3} \times 30 \times 60 = 140.4 \text{ m}^3$$

The area of filtering surface in use,
$$A = 125 \times 10^{-2} \times 70 \times 10^{-2} \times 17 \times 2 = 29.75 \text{ m}^2$$

Now, 1 m$^3$ of dry cake has a mass = 1324 kg

So the volume of solid clay in it

$$= \frac{1324}{2230} = 0.5937 \text{ m}^3$$

Thus, the fractional voidage of the filter cake,

$$e = \frac{1 - 0.5937}{1} = 0.4063$$

So the volume of cake deposited by 1 m³ of filtrate (ref. 2),

$$v = \frac{J\rho}{(1-J)(1-e)\rho_s - Je\rho}$$

$$= \frac{0.0278 \times 998}{(1-0.0278)(1-0.4063) \times 2230 - 0.0278 \times 0.4063 \times 998}$$

$$= 0.021\,75$$

and using equation (8.5), the thickness of filter cake which will have formed after 30 minutes of filtration,

$$l = \frac{0.021\,75 \times 140.4}{29.75} \text{ m}$$

$$= \underline{10.26 \text{ cm}}$$

(iii) Since $\quad K_1 = \dfrac{r\mu v}{A^2} \quad$ (cf. equations (8.1) and (8.2)),

specific cake resistance, $\quad r = \dfrac{K_1 A^2}{\mu v}$

$$= \frac{3.013 \times 10^3 \times 29.75^2}{1.002 \times 10^{-3} \times 0.021\,75} \text{ m}^{-2}$$

$$= 1.224 \times 10^{11} \text{ m}^{-2}$$

Now, since $\quad K_2 = \dfrac{r\mu L}{A} \quad$ (cf. equations (8.1) and (8.2)),

the thickness of cake which is equal in resistance to the resistance of the filter medium,

$$L = \frac{K_2 A}{r\mu}$$

$$= \frac{1.321 \times 10^4 \times 29.75}{1.224 \times 10^{11} \times 1.002 \times 10^{-3}} \text{ m}$$

$$= \underline{3.20 \text{ mm}}$$

148  Problems in fluid flow

(iv) Now by definition,  $r = \dfrac{5S^2(1-e)^2}{e^3}$  (ref. 3)

so  $S^2 = \dfrac{re^3}{5(1-e)^2}$

$= \dfrac{1.224 \times 10^{11} \times 0.4063^3}{5(1-0.4063)^2} \text{ m}^{-2}$

from which,  $S = 6.825 \times 10^4 \text{ m}^{-1}$

Since for spheres,  $S = 6/(\text{particle diameter})$  (ref. 4)

the average diameter of particles in the slurry

$= \dfrac{6}{6.825 \times 10^4} \text{ m}$

$= \underline{87.9 \ \mu\text{m}}$

## 8.1.2 Constant rate filtration, followed by filtration at constant pressure drop, then washing, to illustrate common filtration practice. Incompressible filter cakes

*Problem*

A leaf filter is being used to filter an aqueous slurry which gives rise to an incompressible filter cake. Filtration is being carried out initially at a constant rate of 240 cm³ s⁻¹. After 3 minutes of filtration, the pressure drop developed across the filter = 5.34 kPa, and after 8 minutes, the pressure drop = 9.31 kPa. Calculate:
(i) how long it will take for the pressure drop to reach 15.0 kPa, and
(ii) the volume of filtrate which will have been collected at this stage.
Then if filtration is continued at a constant pressure drop of 15.0 kPa, calculate
(iii) how long it will take to collect a total of 750 dm³ of filtrate. Finally, calculate (iv) how long it will take to wash the filter cake which has accumulated with 10 dm³ of water with a pressure drop of 12 kPa across the filter.

*Solution*

The basic equation by which parameters which affect filtration may be correlated when an incompressible filter cake is formed is (ref. 1):

$$\dfrac{dV}{dt} = \dfrac{A^2(-\Delta P)}{vr\mu(V + LA/v)}$$

from which,  $-\Delta P \equiv K_1 \cdot \dfrac{dV}{dt} \cdot V + K_2 \cdot \dfrac{dV}{dt}$  (8.6)

If filtration is conducted at constant rate,

$$\frac{dV}{dt} = \frac{V}{t}$$

Then, by using this equation to substitute for $V$ in equation (8.6), for constant rate filtration,

$$-\Delta P = K_1 \left(\frac{dV}{dt}\right)^2 \cdot t + K_2 \left(\frac{dV}{dt}\right) \qquad (8.7)$$

Substituting data which obtain after 3 minutes of filtration into equation (8.7)

$$5.34 \times 10^3 = K_1 (240 \times 10^{-6})^2 \times 3 \times 60 + K_2 \times 240 \times 10^{-6} \qquad (8.8)$$

Now substituting data which obtain after 8 minutes,

$$9.31 \times 10^3 = K_1 (240 \times 10^{-6})^2 \times 8 \times 60 + K_2 \times 240 \times 10^{-6} \qquad (8.9)$$

Equations (8.8) and (8.9) are simultaneous equations, from which:

$$K_1 = 2.297 \times 10^8 \text{ kg m}^{-7} \text{s}^{-1}$$

and

$$K_2 = 1.233 \times 10^7 \text{ kg m}^{-4} \text{s}^{-1}$$

(i) Now if filtration is continued at 240 cm³ s⁻¹, substituting in equation (8.7), the time $t$ it will take for the pressure drop to reach 15.0 kPa is given by:

$$15.0 \times 10^3 = 2.297 \times 10^8 (240 \times 10^{-6})^2 \times t + 1.233 \times 10^7 \times 240 \times 10^{-6}$$

from which $t = 910$ s

$$= \underline{15 \text{ min } 10 \text{ s}}$$

(ii) Since filtration will have been conducted at constant rate during this period, the volume of filtrate which will have been collected

$$= 240 \times 10^{-6} \times 910 \text{ m}^3$$

$$= \underline{218.4 \text{ dm}^3}$$

(iii) Equation (8.6) may now be integrated for filtration at constant pressure drop. Hence,

$$t_2 - t_1 = \frac{K_1}{2(-\Delta P)}(V_2^2 - V_1^2) + \frac{K_2}{(-\Delta P)}(V_2 - V_1)$$

Now if $t_1$ is taken as time at the end of the constant rate period, $V_1 = 0.2184$ m³. Hence the time, $t_2$, it will take to collect a total of 750 dm³ of filtrate, by filtering with a constant pressure drop of 15 kPa after the constant rate period, is given by:

$$t_2 - 910 = \frac{2.297 \times 10^8}{2 \times 15.0 \times 10^3}(0.750^2 - 0.2184^2) + \frac{1.233 \times 10^7}{15.0 \times 10^3}(0.750 - 0.2184)$$

From this,
$$t_2 = 5289 \text{ s}$$
$$= \underline{88 \text{ min } 9 \text{ s}}$$

(iv) Since filter cake will not be accumulating during the washing stage, both flow rate through the cake and pressure drop will be constant. Equation (8.6) may be used, therefore, to calculate flow rate $(dV/dt)$ under washing conditions. The value of $V$ required in equation (8.6) is now the total volume of filtrate which will have passed to build up filter cake. Thus:

$$12 \times 10^3 = 2.297 \times 10^8 \times 750 \times 10^{-3} (dV/dt) + 1.233 \times 10^7 (dV/dt)$$

from which,  $\quad dV/dt = 65.00 \text{ cm}^3 \text{ s}^{-1}$

So, the time required to pass 10 dm³ of wash water through the cake at this rate

$$= \frac{10 \times 10^{-3}}{65.00 \times 10^{-6}} \text{ s}$$

$$= 154 \text{ s,}$$

$$= \underline{2 \text{ min } 34 \text{ s}}$$

## 8.1.3 Determination of the characteristics of a filtration system (incompressible filter cake) with a small leaf filter, followed by manipulations to determine the performance of a large plate-and-frame filter, the latter being operated firstly under constant rate conditions, then with a constant pressure drop

*Problem*

A leaf filter, with 650 cm² of filter surface, was used in preliminary tests of the filtering characteristics of an aqueous slurry which gives rise to an incompressible filter cake. The slurry contained a mass fraction of 0.0876 of solid (density = 3470 kg m$^{-3}$) in suspension. When a pressure difference of 12 kPa was established across the filter, 33.80 dm³ of filtrate was obtained after 3 minutes of filtration, and an additional 23.25 dm³ was obtained after a total of 8 minutes of filtration.

A plate-and-frame filter press, with the same type of filter medium, was then used to filter the slurry. 9 frames were used, each 11 cm thick, with filtering surfaces 70 cm square. Filtration was carried out at a constant rate of 16 dm³ s$^{-1}$ until a pressure drop of 15 kPa was established across the filter. Filtration was then continued at 15 kPa until the free surfaces of filter cake in the frames were separated by 1 cm. Calculate:
(i) the total volume of filtrate which was collected,
(ii) the total time taken for the operation,

(iii) the average diameter of particles in the slurry (assuming that they were spherical), and
(iv) the thickness of filter cake which has a resistance equal to the resistance of the filter medium.

Density of water = 999 kg m$^{-3}$
Viscosity of water = 1.12 × 10$^{-3}$ kg m$^{-1}$ s$^{-1}$
Density of dry filter cake = 1922 kg m$^{-3}$

*Solution*

When an incompressible filter cake forms, the parameters which affect filtration, when filtration is carried out at constant pressure drop, are correlated by the equation (ref. 5):

$$t_2 - t_1 = \frac{K_1}{2(-\Delta P)}(V_2^2 - V_1^2) + \frac{K_2}{(-\Delta P)}(V_2 - V_1) \quad (8.10)$$

in which
$$K_1 = \frac{r\mu v}{A^2} \quad (8.11)$$

and
$$K_2 = \frac{r\mu L}{A} \quad (8.12)$$

Subscripts s and p will now be used to signify $K_1$, $K_2$ and $A$ values which pertain to the small leaf filter and the plate-and-frame filter systems respectively.

Where filtration is carried out at constant pressure drop from the outset, equation (8.10) reduces to:

$$t = \frac{K_1}{2(-\Delta P)}V^2 + \frac{K_2}{(-\Delta P)}V \quad (8.13)$$

Substituting data which obtain after 3 minutes of filtration with the leaf filter,

$$3 \times 60 = \frac{K_{1,s}}{2 \times 12 \times 10^3}(33.80 \times 10^{-3})^2 + \frac{K_{2,s}}{12 \times 10^3}(33.80 \times 10^{-3}) \quad (8.14)$$

After 8 minutes of operation,

$$8 \times 60 = \frac{K_{1,s}}{2 \times 12 \times 10^3}[(33.80 + 23.25) \times 10^{-3}]^2$$

$$+ \frac{K_{2,s}}{12 \times 10^3}(33.80 + 23.25) \times 10^{-3} \quad (8.15)$$

Equations (8.14) and (8.15) are simultaneous equations,
from which $\quad K_{1,s} = 3.189 \times 10^9$ kg m$^{-7}$ s$^{-1}$
and $\quad K_{2,s} = 1.002 \times 10^7$ kg m$^{-4}$ s$^{-1}$

## 152 Problems in fluid flow

Now $K_1$ and $K_2$ values for the two systems depend on the area available for filtration.

$$A_p = (70 \times 10^{-2})^2 \times 9 \times 2 = 8.82 \text{ m}^2$$
$$A_s = 650 \times 10^{-4} \text{ m}^2$$

Slurry characteristics remain constant. So using equation (8.11),

$$K_{1,p} = \frac{r\mu v}{A_p^2}$$

$$= \frac{r\mu v}{A_s^2} \cdot \frac{A_s^2}{A_p^2}$$

$$= K_{1,s} \cdot \frac{A_s^2}{A_p^2}$$

$$= \frac{3.189 \times 10^9 (650 \times 10^{-4})^2}{8.82^2} \text{ kg m}^{-7}\text{s}^{-1}$$

$$= 1.732 \times 10^5 \text{ kg m}^{-7}\text{s}^{-1}$$

Using equation (8.12),

$$K_{2,p} = \frac{r\mu L}{A_p}$$

$$= \frac{r\mu L}{A_s} \cdot \frac{A_s}{A_p}$$

$$= K_{2,s} \cdot \frac{A_s}{A_p}$$

$$= \frac{1.002 \times 10^7 \times 650 \times 10^{-4}}{8.82} \text{ kg m}^{-4}\text{s}^{-1}$$

$$= 7.384 \times 10^4 \text{ kg m}^{-4}\text{s}^{-1}$$

When filtration is carried out at constant rate,

$$-\Delta P = K_1 \left(\frac{dV}{dt}\right)^2 t + K_2 \frac{dV}{dt} \qquad (8.2)$$

Substituting data for the constant rate period,

$$15 \times 10^3 = 1.732 \times 10^5 (16 \times 10^{-3})^2 t + 7.384 \times 10^4 \times 16 \times 10^{-3}$$

from which, the time required for this stage,

$$t = 311.7 \text{ s}$$

The volume of filtrate which would be collected during the constant rate stage

$$= 16 \times 10^{-3} \times 311.7 = 4.987 \text{ m}^3$$

Now when the free surfaces of filter cake are separated by 1 cm, thickness of filter cake $= \dfrac{11-1}{2} = 5.0$ cm

Also, the volume of filter cake deposited by passage of 1 m³ of filtrate (ref. 2),

$$v = \frac{J\rho}{(1-J)(1-e)\rho_s - Je\rho}$$

From data given, voidage of the filter cake,

$$e = \frac{1 - 1922/3470}{1} = 0.4461$$

So, $\quad v = \dfrac{0.0876 \times 999}{(1-0.0876)(1-0.4461) \times 3470 - 0.0876 \times 0.4461 \times 999}$,

$= 0.051\,04$

Also since $\quad\quad\quad\quad v \cdot V = A \cdot l \quad\quad\quad\quad$ (ref. 2)

the total volume of filtrate which will have been collected when the filter cake is 5.0 cm thick,

$$V = \frac{8.82 \times 5.0 \times 10^{-2}}{0.051\,04}\text{ m}^3$$

$= \underline{8.640\text{ m}^3}$

Now substituting data into equation (8.10),

$$t_2 - 311.7 = \frac{1.732 \times 10^5}{2 \times 15 \times 10^3}(8.640^2 - 4.987^2) + \frac{7.384 \times 10^4}{15 \times 10^3}(8.640 - 4.987)$$

from which $\quad\quad\quad\quad t_2 = 617$ s

$= 10$ min 17 s

i.e. the total time required for the whole operation $= \underline{10 \text{ min } 17\text{ s}}$

Now $\quad\quad\quad\quad K_{1,p} = \dfrac{r\mu v}{A_p^2}$

$= 1.732 \times 10^5 \text{ kg m}^{-7}\text{s}^{-1}$

So, specific cake resistance,

$$r = \frac{1.732 \times 10^5 \times 8.82^2}{1.12 \times 10^{-3} \times 0.051\,04}\text{ m}^{-2}$$

$= 2.357 \times 10^{11} \text{ m}^{-2}$

But $\quad\quad\quad\quad r = \dfrac{5S^2(1-e)^2}{e^3} \quad\quad\quad\quad$ (ref. 3)

So
$$S^2 = \frac{2.357 \times 10^{11} \times 0.4461^3}{5(1-0.4461)^2} \, m^{-2}$$

from which $\quad S = 1.168 \times 10^5 \, m^{-1}$

But for spherical particles,

$$S = 6/(\text{particle diameter}) \qquad \text{(ref. 4)}$$

So, the average diameter of particles in the slurry

$$= \frac{6}{1.168 \times 10^5} \, m$$

$$= \underline{51.4 \, \mu m}$$

Finally, $\quad K_{2,p} = \dfrac{r\mu L}{A_p}$

$$= 7.384 \times 10^4 \, kg \, m^{-4} \, s^{-1}$$

So
$$L = \frac{7.384 \times 10^4 \times 8.82}{2.357 \times 10^{11} \times 1.12 \times 10^{-3}} \, m$$

$$= 2.47 \, mm$$

i.e. the thickness of filter cake which has a resistance equal to the resistance of the filter medium = 2.47 mm.

## 8.1.4 Constant pressure drop filtration of a suspension which gives rise to a compressible filter cake

*Problem*

In test runs, a suspension, which gives rise to a compressible filter cake, was filtered on a press which afforded 1.86 m² of filtering surface. With a pressure drop of 45 kPa over the filter, 5.21 dm³ of filtrate was obtained after 3 minutes of filtration, and a total of 17.84 dm³ of filtrate was obtained after 12 minutes. In a second run with a pressure drop of 85 kPa across the filter, 10.57 dm³ of filtrate was obtained after 5 minutes.

If the same suspension is now filtered on the same filter press with a pressure drop of 78 kPa, calculate the volume of filtrate which will have been collected after filtering for 90 minutes. Calculate also the volume of filtrate which would be collected by filtering for 90 minutes with a pressure drop of 45 kPa. The suspension contains 11.8 kg of solids in 1 m³ of liquid.

Viscosity of the liquid = $1.29 \times 10^{-3} \, kg \, m^{-1} \, s^{-1}$

*Filtration* 155

*Solution*

When a compressible filter cake forms, and resistance to flow derives from interaction of fluid with filter cake alone, flow rate of filtrate; (ref. 6),

$$\frac{dV}{dt} = \frac{A^2(-\Delta P)}{V\mu c \bar{r}} \qquad (8.16)$$

in which average specific cake resistance (ref. 6),

$$\bar{r} = \bar{r}'(1-n')\Delta P^{n'} \qquad (8.17)$$

($\bar{r}'$ and $n'$ are constants)
If filtration is carried out with a constant pressure drop throughout, equation (8.16) may be integrated to give:

$$t = \frac{\mu c \bar{r}}{2A^2(-\Delta P)} \cdot V^2 \qquad (8.18)$$

Now if resistance of the filter medium is also taken into account, by reasoning similar to that which obtains for incompressible filter cakes (ref. 1), and for filtration again with a constant pressure drop, the equation which obtains to replace equation (8.18), is:

$$t = \frac{\bar{r}\mu c}{2A^2(-\Delta P)} \cdot V^2 + \frac{k}{(-\Delta P)} \cdot V \qquad (8.19)$$

($k$ is a constant)
Now $\bar{r}$ has a value which depends on pressure drop (see equation (8.17)). So, substituting data for the $-\Delta P = 45$ kPa system into equation (8.19), gives firstly:

$$3 \times 60 = \frac{\bar{r}_{45} \times 1.29 \times 10^{-3} \times 11.8 (5.21 \times 10^{-3})^2}{2 \times 1.86^2 \times 45 \times 10^3} + \frac{k \times 5.21 \times 10^{-3}}{45 \times 10^3} \qquad (8.20)$$

then:

$$12 \times 60 = \frac{\bar{r}_{45} \times 1.29 \times 10^{-3} \times 11.8 (17.84 \times 10^{-3})^2}{2 \times 1.86^2 \times 45 \times 10^3} + \frac{k \times 17.84 \times 10^{-3}}{45 \times 10^3} \qquad (8.21)$$

Equations (8.20) and (8.21) are simultaneous equations, from which

$$\bar{r}_{45} = 9.424 \times 10^{12} \text{ m kg}^{-1}$$

and

$$k = 1.446 \times 10^9 \text{ kg m}^{-4} \text{ s}^{-1}$$

156  *Problems in fluid flow*

Now substituting this value for $k$ and data for $-\Delta P = 85$ kPa into equation (8.19),

$$5 \times 60 = \frac{\bar{r}_{85} \times 1.29 \times 10^{-3} \times 11.8(10.57 \times 10^{-3})^2}{2 \times 1.86^2 \times 85 \times 10^3}$$

$$+ \frac{1.446 \times 10^9 \times 10.57 \times 10^{-3}}{85 \times 10^3}$$

From this, $\quad \bar{r}_{85} = 4.156 \times 10^{13}$ m kg$^{-1}$

Now substituting $\bar{r}$ and corresponding $-\Delta P$ values into equation (8.17),

$$\bar{r}_{45} = 9.424 \times 10^{12} = \bar{r}'(1-n')(45 \times 10^3)^{n'} \tag{8.22}$$

and $\quad \bar{r}_{85} = 4.156 \times 10^{13} = \bar{r}'(1-n')(85 \times 10^3)^{n'} \tag{8.23}$

Dividing equation (8.22) by equation (8.23) and simplifying,

$$\frac{9.424 \times 10^{12}}{4.156 \times 10^{13}} = \left(\frac{45}{85}\right)^{n'}$$

from which $\quad n' = 2.333$

Now substituting $n' = 2.333$ into equation (8.22),

$$9.424 \times 10^{12} = \bar{r}'(1 - 2.333)(45 \times 10^3)^{2.333}$$

Hence, $\quad \bar{r}' = -98.505$ (S.I. units)

Using equation (8.17) again, for a pressure drop of 78 kPa, average specific cake resistance,

$$\bar{r}_{78} = -98.505(1 - 2.333)(78 \times 10^3)^{2.333} \text{ m kg}^{-1}$$

$$= 3.401 \times 10^{13} \text{ m kg}^{-1}$$

Now substituting data into equation (8.19), after 90 minutes of filtration,

$$90 \times 60 = \frac{3.401 \times 10^{13} \times 1.29 \times 10^{-3} \times 11.8}{2 \times 1.86^2 \times 78 \times 10^3} V^2 + \frac{1.446 \times 10^9}{78 \times 10^3} V$$

from which $\quad V = 65.99$ dm$^3$

i.e. the volume of filtrate which will have been collected after 90 minutes of filtration with a pressure drop of 78 kPa = 66.0 dm$^3$.

Similarly, by substituting data relevant to a 45 kPa pressure-drop system, after 90 minutes of filtration,

$$90 \times 60 = \frac{9.424 \times 10^{12} \times 1.29 \times 10^{-3} \times 11.8}{2 \times 1.86^2 \times 45 \times 10^3} V^2 + \frac{1.446 \times 10^9}{45 \times 10^3} V$$

from which, $\quad V = 78.87$ dm$^3$

i.e. the volume of filtrate which would be collected after 90 minutes of filtration with a pressure drop of 45 kPa = 78.9 dm³.

### 8.1.5 Filtration on a rotary drum filter

*Problem*

An aqueous slurry, which contains 0.154 kg of solid barium carbonate (density = 4430 kg m$^{-3}$, average particle diameter (assuming that particles are spherical) = 34 µm) in each kilogram of suspension, is being filtered on a rotary drum filter (diameter = 1.2 m, length = 2.6 m). The carbonate forms an incompressible cake with a fractional voidage of 0.415. If a vacuum pump can maintain a pressure differential across the filtering surface of 34.8 kPa, and the drum rotates at 0.5 rev min$^{-1}$, with 30% of its cylindrical surface being immersed in slurry at any one time, calculate:
(i) the thickness of filter cake which will form on the drum,
(ii) the rate (in kg h$^{-1}$) at which wet cake may be scraped from the surface of the drum, and
(iii) the rate (in kg h$^{-1}$) at which slurry may be treated by the filter.

Viscosity of water = $1.17 \times 10^{-3}$ kg m$^{-1}$ s$^{-1}$
Density of water = 999 kg m$^{-3}$
Filter medium resistance is equivalent to 2.35 mm of cake

*Solution*

The basic filtration equation which applies when resistance to flow derives from both filter medium and an incompressible filter cake is (ref. 1):

$$-\Delta P = \frac{r\mu v}{A^2} \cdot \frac{dV}{dt} \cdot V + \frac{r\mu L}{A} \cdot \frac{dV}{dt}$$

If an operation is carried out with constant pressure drop, this equation becomes:

$$dt = \frac{r\mu v}{A^2(-\Delta P)} \cdot V dV + \frac{r\mu L}{A(-\Delta P)} \cdot dV \quad (8.24)$$

Now, $\qquad v \cdot V = A \cdot l \qquad$ (ref. 2)

so, $\qquad V = \frac{A}{v} \cdot l \qquad (8.25)$

and since both $A$ and $v$ are constant,

$$dV = \frac{A}{v} \cdot dl \quad (8.26)$$

## 158   Problems in fluid flow

Now, using equations (8.25) and (8.26) to substitute in equation (8.24), after simplifying,

$$\mathrm{d}t = \frac{r\mu}{v(-\Delta P)} \cdot l\,\mathrm{d}l + \frac{r\mu L}{v(-\Delta P)} \cdot \mathrm{d}l$$

Integrating for the time an element of surface is immersed in slurry,

$$\int_0^t \mathrm{d}t = \frac{r\mu}{v(-\Delta P)} \int_0^l l\,\mathrm{d}l + \frac{r\mu L}{v(-\Delta P)} \int_0^l \mathrm{d}l$$

so that in order to build up a cake of thickness $l$ with a constant pressure drop, the time required

$$t = \frac{r\mu}{2v(-\Delta P)} l^2 + \frac{r\mu L}{v(-\Delta P)} l \qquad (8.27)$$

(i) Now during each revolution, the time for which each element of drum surface is collecting cake,

$$t = \frac{60 \times 0.3}{0.5} = 36 \text{ s}$$

Also, for spherical particles,

$$S = 6/d \qquad \text{(ref. 4)}$$

$$= \frac{6}{34 \times 10^{-6}} = 1.765 \times 10^5 \text{ m}^{-1}$$

So, specific cake resistance,

$$r = \frac{5S^2(1-e)^2}{e^3} \qquad \text{(ref. 3)}$$

$$= \frac{5(1.765 \times 10^5)^2 (1-0.415)^2}{0.415^3} \text{ m}^{-2}$$

$$= 7.458 \times 10^{11} \text{ m}^{-2}$$

Also,  $$v = \frac{J\rho}{(1-J)(1-e)\rho_s - Je\rho} \qquad \text{(ref. 2)}$$

$$= \frac{0.154 \times 999}{(1-0.154)(1-0.415) \times 4430 - 0.154 \times 0.415 \times 999}$$

$$= 0.072\,28$$

Now, substituting data into equation (8.27),

$$36 = \frac{7.458 \times 10^{11} \times 1.17 \times 10^{-3}}{2 \times 0.072\,28 \times 34.8 \times 10^3} l^2$$

$$+ \frac{7.458 \times 10^{11} \times 1.17 \times 10^{-3} \times 2.35 \times 10^{-3}}{0.072\,28 \times 34.8 \times 10^3} l$$

from which, $\quad l^2 + 4.6986 \times 10^{-3} l - 2.0749 \times 10^{-4} = 0$

so that $\quad\quad\quad\quad\quad\quad\quad\quad\quad\quad l = 12.25$ mm

i.e., thickness of filter cake which will form on the drum = 12.25 mm

(ii) The velocity with which the drum surface passes the scraper knife

$$= \frac{\pi \times 1.2 \times 0.5}{60} \text{ m s}^{-1}$$

$$= 0.03142 \text{ m s}^{-1}$$

So the volumetric rate at which cake is removed by the scraper knife

$$= 0.03142 \times 2.6 \times 12.25 \times 10^{-3} \text{ m}^3 \text{s}^{-1}$$

$$= 1.001 \times 10^{-3} \text{ m}^3 \text{s}^{-1}$$

Thus, the mass rate at which solid is removed in cake

$$= 1.001 \times 10^{-3} (1 - 0.415) \times 4430 \text{ kg s}^{-1}$$

$$= 2.594 \text{ kg s}^{-1}$$

The mass rate at which water is removed with cake by the scraper knife

$$= 1.001 \times 10^{-3} \times 0.415 \times 999 \text{ kg s}^{-1}$$

$$= 0.415 \text{ kg s}^{-1}$$

So the rate at which wet cake will be scraped off the surface of the drum

$$= 2.594 + 0.415 = 3.009 \text{ kg s}^{-1}$$

$$= 1.083 \times 10^4 \text{ kg h}^{-1}$$

(iii) The rate at which solids are scraped off the drum

$$= 2.594 \text{ kg s}^{-1}$$

But 0.154 kg of solid is contained in each kilogram of suspension. So the rate at which slurry may be treated

$$= \frac{2.594}{0.154} \text{ kg s}^{-1}$$

$$= 6.064 \times 10^4 \text{ kg h}^{-1}$$

## 8.1.6 Filtration in a centrifugal filter

*Problem*

An aqueous suspension of sucrose crystals (density = 1581 kg m$^{-3}$), which gives rise to an incompressible filter cake, is to be filtered in a centrifugal filter.

160   Problems in fluid flow

The sucrose crystals are cubes, with an average side length = 85 μm, which pack to form a cake of voidage = 0.435, and which may be considered to make point contact only with neighbouring crystals in the filter cake. The centrifuge basket is 2.5 m deep, has a diameter of 3 m, and will rotate at 325 rev min$^{-1}$. The diameter of the inner surface of the suspension can be held constant in the basket at 2.10 m. If 1 kg of suspension contains 0.097 kg of crystals, and filter cloth resistance is equivalent to a layer of cake 4.8 mm thick, calculate:
(i) how long it will take to collect 3500 kg of sucrose crystals in a filter cake, and
(ii) the volume of mother liquor which will separate as filtrate.

Viscosity of mother liquor = 0.224 kg m$^{-1}$ s$^{-1}$
Density of mother liquor = 1328 kg m$^{-3}$
Kozeny's constant = 5.0

*Solution*

(i) The equation which correlates parameters which affect centrifugal filtration is (ref. 7):

$$(b^2 - b'^2)\left(1 + \frac{2L}{b}\right) + 2b'^2 \ln\frac{b'}{b} = \frac{2vt\rho\omega^2}{r\mu}(b^2 - x^2) \quad (8.28)$$

For this equation,

$$b = 3 \times 0.5 = 1.5 \text{ m}$$

$$x = 2.10 \times 0.5 = 1.05 \text{ m}$$

Now the volume of filter cake which will be collected

$$= \pi b^2 H - \pi b'^2 H$$

$$= \pi H(b^2 - b'^2)$$

If the cake has a voidage, $e$, the mass of solids in the filter cake

$$= \pi H(b^2 - b'^2)(1-e)\rho_s$$

So substituting data into this equation, the inner radius, $b'$, to which cake must be built up may be obtained from:

$$3500 = \pi \times 2.5(1.5^2 - b'^2)(1 - 0.435) \times 1581$$

Hence,      $b' = 1.323$ m

Also,
$$v = \frac{J\rho}{(1-J)(1-e)\rho_s - Je\rho} \quad \text{(ref. 2)}$$

$$= \frac{0.097 \times 1328}{(1-0.097)(1-0.435) \times 1581 - 0.097 \times 0.435 \times 1328}$$

$$= 0.1716$$

*Filtration* 161

Next, $\omega = 325 \text{ rev min}^{-1}$

$$= \frac{325 \times 2 \times \pi}{60} = 34.03 \text{ rad s}^{-1}$$

Also, average surface area of a crystal $= 6 \times (85 \times 10^{-6})^2 = 4.335 \times 10^{-8} \text{ m}^2$

The average volume of a crystal $= (85 \times 10^{-6})^3 = 6.141 \times 10^{-13} \text{ m}^3$

So, $$S = \frac{4.335 \times 10^{-8}}{6.141 \times 10^{-13}} = 7.059 \times 10^4 \text{ m}^{-1}$$

and specific cake resistance,

$$r = \frac{K''S^2(1-e)^2}{e^3} \quad \text{(ref. 3 and 8)}$$

$$= \frac{5.0(7.059 \times 10^4)^2(1-0.435)^2}{0.435^3} \text{ m}^{-2}$$

$$= 9.662 \times 10^{10} \text{ m}^{-2}$$

Finally, the time, $t$, it will take to collect 3500 kg of sucrose crystals in a filter cake may be obtained by substituting data given and derived into equation (8.28). Thus:

$$(1.5^2 - 1.323^2)\left(1 + \frac{2 \times 4.8 \times 10^{-3}}{1.5}\right) + 2 \times 1.323^2 \times \ln\left(\frac{1.323}{1.5}\right)$$

$$= \frac{2 \times 0.1716 \times 1328 \times 34.03^2 \times (1.5^2 - 1.05^2)}{9.662 \times 10^{10} \times 0.224} t$$

from which $t = 2263 \text{ s}$

$$= \underline{37 \text{ min } 43 \text{ s}}$$

(ii) Now the volume of cake which will separate

$$= (\pi b^2 - \pi b'^2)H$$

$$= \pi \times 2.5(1.5^2 - 1.323^2) = 3.924 \text{ m}^3$$

Since the voidage of the cake $= 0.435$
the volume of liquid which will be retained in the cake

$$= 3.924 \times 0.435 = 1.707 \text{ m}^3$$

But also, the volume of cake (ref. 2), $= v \cdot V$
So the volume of mother liquor associated initially with solids which form the cake,

$$V = \frac{3.924}{0.1716} = 22.867 \text{ m}^3$$

Allowing for the mother liquor retained by the cake, the volume of liquor which will separate as filtrate

$$= 22.867 - 1.707 \text{ m}^3$$
$$= \underline{21.16 \text{ m}^3}$$

## 8.2 Student Exercises

1  A slurry of gypsum (which forms an incompressible filter cake) in water, containing a mass fraction of $CaSO_4 = 0.0133$, was filtered in a plate-and-frame filter press, 30 cm square, with 10 frames in operation. Filtration was carried out at a constant rate of 8 dm$^3$ s$^{-1}$. After 1 minute of filtration, the pressure drop across the filter was 5.27 kPa, and after 5 minutes, it was 20.34 kPa. What was the pressure drop after 20 minutes of filtration, and how thick was the filter cake at this stage? Also, find the specific cake resistance, $r$, the average diameter of the particles (assuming that they are spherical), and the thickness of cake equivalent in resistance to the resistance of the filter medium.

Density of water = 999 kg m$^{-3}$
Density of gypsum = 2610 kg m$^{-3}$
Density of the dry cake = 1500 kg m$^{-3}$
Viscosity of the filtrate = $1.05 \times 10^{-3}$ kg m$^{-1}$ s$^{-1}$

2  A leaf filter was used to make preliminary tests on the filtration characteristics of an aqueous slurry which gave rise to an incompressible filter cake. When filtration was carried out at a constant rate of 25 cm$^3$ s$^{-1}$, after 1 minute of filtration, the pressure drop across the filter was found to be 48.5 kPa, and after 10 minutes of filtration, it was 284 kPa.

The same filter was then used to filter the same suspension at a constant rate of 40 cm$^3$ s$^{-1}$ until a pressure drop of 350 kPa was established across the filter. Filtration was then continued, with the pressure drop held constant at this value, until the total volume of filtrate collected was 100 dm$^3$. Calculate:
(i) how long it took for this whole operation, and
(ii) how long it took to wash the accumulated filter cake with 8 dm$^3$ of water with a pressure drop of 300 kPa across the filter.

3  A leaf filter, with 500 cm$^2$ of filtering surface, was used in preliminary tests of the filtration characteristics of a slurry of barium sulphate (which forms an incompressible filter cake), in water. The slurry contained a mass fraction of 0.335 of solid in suspension. When a pressure difference of 65 kPa was established across the filter, 248 cm$^3$ of filtrate was recovered after 5 minutes of operation, and a total of 413 cm$^3$ of filtrate was obtained after 10 minutes.

A plate-and-frame filter press, with the same type of filter medium, will now be used to filter the slurry. 12 frames, each 30 cm square, will be used. Filtration

Filtration  163

will be carried out at a constant rate of 100 cm³ s⁻¹ until a pressure drop of 400 kPa has been established across the filter. Filtration will then be continued with this pressure drop for a further 30 minutes. Finally, the filter cake will be washed with water for 10 minutes with a pressure differential of 250 kPa. Calculate:
(i) the total volume of filtrate which will be collected,
(ii) the volume of wash water required,
(iii) the thickness of filter cake which will form,
(iv) the average diameter of the particles of barium sulphate (assuming that they are spherical), and
(v) the thickness of cake equivalent in resistance to the resistance of the filter medium.

Density of barium sulphate = 4500 kg m⁻³
Density of the dry filter cake = 2770 kg m⁻³
Density of water = 1000 kg m⁻³
Viscosity of water = $1.42 \times 10^{-3}$ kg m⁻¹ s⁻¹

4  When a suspension of a compressible sludge was filtered with a pressure drop of 100 kPa across a filter press, with an area of 2.16 m² available for filtration, 8.84 dm³ of filtrate was obtained after 2 minutes of filtration, and 32.1 dm³ of filtrate was obtained after 10 minutes. In a further test with a pressure drop of 400 kPa, 21.2 dm³ of filtrate was obtained after 2 minutes of filtration.

Find how long it will take to obtain 150 dm³ of filtrate if the suspension is filtered on the same filter press with a pressure drop of 250 kPa across the assembly. The suspension contains 514 g of solids dispersed in 1 dm³ of water.

Viscosity of water = $1.42 \times 10^{-3}$ kg m⁻¹ s⁻¹

5  An aqueous slurry of benzoic acid crystals (density = 1266 kg m⁻³) forms an incompressible cake (fractional voidage = 0.508) 14.5 mm thick on the cylindrical surface of a rotary drum filter (diameter = 80 cm, length = 120 cm). The filter drum turns at a rate of 1 rev min⁻¹, and has 24% of its cylindrical surface submerged in slurry at any one time. The pressure drop which can be maintained across the filtering surface = 13.5 kPa. If the benzoic acid crystals may be considered cylindrical (average diameter = 120 μm, average length = 310 μm), make only point contacts with other crystals in the filter cake, and form a cake characterised by a Kozeny constant = 5.4, calculate:
(i) the concentration of the slurry (in g benzoic acid per kg of slurry), and
(ii) the volumetric flow rate of filtrate.

Viscosity of the aqueous phase = $2.15 \times 10^{-3}$ kg m⁻¹ s⁻¹
Density of the aqueous phase = 1040 kg m⁻³
The filter medium resistance is equivalent to a layer of cake 4.6 mm thick.

164  *Problems in fluid flow*

6   Sucrose crystals (density = 1581 kg m$^{-3}$), which give rise to an incompressible filter cake, are to be filtered from suspension in an aqueous mother liquor in a centrifugal filter. The crystals are cubes, with an average side length = 110 $\mu$m, which pack to form a cake of voidage 0.462, and which may be considered to make point contact only with neighbouring crystals in the filter cake. The centrifuge basket is 50 cm deep, has a diameter = 58 cm, and will rotate at 1450 rev min$^{-1}$. The diameter of the inner surface of the suspension in the centrifuge basket can be held constant at 24 cm. If 1 kg of suspension contains 43 g of crystals, and the filter cloth resistance is equivalent to a layer of cake 7.5 mm thick, calculate:
(i) the mass (in kg) of crystals which will collect in a filter cake after 12 minutes of filtration, and
(ii) the volume of mother liquor which will separate as filtrate.

Viscosity of mother liquor = 0.228 kg m$^{-1}$ s$^{-1}$
Density of mother liquor = 1331 kg m$^{-3}$
Kozeny's constant = 5.0

## 8.3   References

All of the following references are to be found in:

J. M. Coulson and J. F. Richardson, Chemical Engineering, Vol. 2, 3rd edition, Pergamon, 1980

1. page 326
2. page 324
3. page 323
4. page 126
5. page 327
6. page 330
7. page 364
8. page 129

## 8.4   Notation

| Symbol | Description | Unit |
|---|---|---|
| $A$ | area of filtering surface | m$^2$ |
| $A_p$ | area of filtering surface in a plate-and-frame filter | m$^2$ |
| $A_s$ | area of filtering surface in a small leaf filter | m$^2$ |
| $b$ | radius of the basket of a centrifugal filter | m |
| $b'$ | radius of the free surface of the filter cake in a centrifugal filter system | m |
| $c$ | mass of solids contained in unit volume of filtrate | kg m$^{-3}$ |
| $d$ | average diameter of particles | m |
| $e$ | fractional voidage of the filter cake | — |
| $H$ | height of the basket of a centrifugal filter | m |
| $J$ | mass of solid contained in unit mass of suspension | — |
| $k$ | a constant | kg m$^{-4}$ s$^{-1}$ |
| $K''$ | Kozeny's constant | — |

## Filtration  165

| Symbol | Description | Unit |
|---|---|---|
| $K_1$ | $r\mu v/A^2$ | $\text{kg m}^{-7}\text{s}^{-1}$ |
| $K_{1,p}$ | $K_1$ for a plate-and-frame filter system | $\text{kg m}^{-7}\text{s}^{-1}$ |
| $K_{1,s}$ | $K_1$ for a small leaf-filter system | $\text{kg m}^{-7}\text{s}^{-1}$ |
| $K_2$ | $r\mu L/A$ | $\text{kg m}^{-4}\text{s}^{-1}$ |
| $K_{2,p}$ | $K_2$ for a plate-and-frame filter system | $\text{kg m}^{-4}\text{s}^{-1}$ |
| $K_{2,s}$ | $K_2$ for a small leaf-filter system | $\text{kg m}^{-4}\text{s}^{-1}$ |
| $l$ | thickness of the filter cake | m |
| $L$ | thickness of cake equivalent in resistance to the resistance of the filter medium | m |
| $n'$ | a constant | — |
| $r$ | specific cake resistance | $\text{m}^{-2}$ |
| $\bar{r}$ | average specific cake resistance $= \bar{r}'(1-n')\Delta P^{n'}$ | $\text{m kg}^{-1}$ |
| $\bar{r}_{45}$ | $\bar{r}$ for a filter cake with a 45 kPa pressure drop across it | $\text{m kg}^{-1}$ |
| $\bar{r}_{78}$ | $\bar{r}$ for a filter cake with a 78 kPa pressure drop across it | $\text{m kg}^{-1}$ |
| $\bar{r}_{85}$ | $\bar{r}$ for a filter cake with a 85 kPa pressure drop across it | $\text{m kg}^{-1}$ |
| $\bar{r}'$ | a constant | — |
| $S$ | average surface area of particles per unit volume of them | $\text{m}^{-1}$ |
| $t$ | time | s |
| $t_1$ | time at the beginning of a period of filtration | s |
| $t_2$ | time at the end of a period of filtration | s |
| $v$ | volume of filter cake deposited by passage of a unit volume of filtrate | |
| $V$ | volume of filtrate passed in time $t$ | $\text{m}^3$ |
| $V_1$ | volume of filtrate passed at time $t_1$ | $\text{m}^3$ |
| $V_2$ | volume of filtrate passed at time $t_2$ | $\text{m}^3$ |
| $x$ | radius of the free surface of liquid suspension in a centrifugal filter system | m |
| $\Delta P$ | pressure difference across a filter system | Pa |
| $\mu$ | viscosity of filtrate | $\text{kg m}^{-1}\text{s}^{-1}$ |
| $\rho$ | density of filtrate | $\text{kg m}^{-3}$ |
| $\rho_s$ | density of solids | $\text{kg m}^{-3}$ |
| $\omega$ | angular velocity of the basket as it rotates in a centrifugal filter system | $\text{rad s}^{-1}$ |

# 9 Forces on bodies immersed in fluids

## 9.1 Worked Examples

### 9.1.1 Drag forces and drag coefficients

*Problem*

(i) A spherical storage tank which has a diameter of 3.0 m is to be erected above the ground. Calculate the drag force experienced by this tank when the wind blows horizontally at 50 km h$^{-1}$.
(ii) What is the drag force on a cube-shaped tank immersed in an air stream with one side normal to the direction of flow if the length of each side of the cube is 0.25 m, and the air flows at 2 m s$^{-1}$.

Density of the air = 1.2 kg m$^{-3}$
Viscosity of the air = 1.85 × 10$^{-5}$ kg m$^{-1}$ s$^{-1}$

*Solution*

(i) The equation for the drag force on a solid body immersed in a flowing fluid (ref. 1, 2, 3) may be obtained from

$$\frac{F_D}{A_p} = \frac{C_D \rho u^2}{2} \qquad (9.1)$$

The value of the drag coefficient, $C_D$, for use in equation (9.1) can be found by first determining the Reynolds number, defined by

$$Re = \frac{du\rho}{\mu} \qquad (9.2)$$

and consulting a chart in which the drag coefficient is plotted as a function of this Reynolds number (ref. 4, 5, 6). In some texts (e.g. ref. 7) a drag coefficient is obtained from the equation

$$\frac{F_D}{A_p} = C_D \rho u^2 \qquad (9.3)$$

Equation (9.3) leads to a drag coefficient which is numerically one half of that defined by equation (9.1). Throughout this and later chapters, equation (9.1) will be used as the defining equation relating the drag force, $F_D$, and the drag coefficient, $C_D$.

Now the sphere has a diameter of 3 m, and the air velocity is

$$= \frac{50 \times 10^3}{3600} \, \text{m s}^{-1}$$

$$= 13.89 \, \text{m s}^{-1}$$

So the Reynolds number, $Re$, is given by

$$Re = \frac{3 \times 13.89 \times 1.2}{1.85 \times 10^{-5}}$$

$$= 2.70 \times 10^6$$

From a chart of drag coefficient versus Reynolds number (ref. 4, 5, 6) for a sphere, the drag coefficient, $C_D$, is 0.2.

Now for a spherical body the projected area $A_p = \dfrac{\pi d^2}{4}$

Thus from equation (9.1) we have the drag force,

$$F_D = A_p C_D \rho u^2 / 2$$

$$= \left(\frac{\pi \times 3^2}{4}\right) \frac{0.2 \times 1.2 \times 13.9^2}{2} \, \text{N}$$

$$= \underline{163.9 \, \text{N}}$$

(ii) For a cube with each side of length $L$, the Reynolds number is defined as

$$Re = \frac{Lu\rho}{\mu}$$

$$= \frac{0.25 \times 2 \times 1.2}{1.85 \times 10^{-5}} = 3.24 \times 10^4$$

To determine the drag coefficient corresponding to this Reynolds number the sphericity must be calculated. Now the sphericity, $\psi$, is defined thus (ref. 8, 9)

$$\psi = \frac{\begin{pmatrix} \text{surface area of sphere of volume} \\ \text{equal to that of the solid} \end{pmatrix}}{\text{the surface area of the solid}} \quad (9.4)$$

The volume of the cube is $L^3$. For a sphere of this volume and with a radius $r$,

$$L^3 = \frac{4}{3}\pi r^3$$

whence
$$r = \left(\frac{3}{4\pi}\right)^{1/3} L$$

The surface area of this sphere
$$= 4\pi r^2 = 4\pi \left(\frac{3}{4\pi}\right)^{2/3} L^2$$

The surface area of the cube $= 6 L^2$

Thus from equation (9.4)
$$\psi = \frac{4\pi \left(\frac{3}{4\pi}\right)^{2/3}}{6}$$
$$= 0.806$$

From the Reynolds number of $3.24 \times 10^4$ and the sphericity of 0.806, the drag coefficient, $C_D$, can be read from a chart of drag coefficient versus Reynolds number (ref. 5).
$$C_D = 2 \text{ (approximately)}$$

Furthermore, the projected area, $A_p$, of the cube $= L^2 = 0.25^2 \text{ m}^2 = 0.0625 \text{ m}^2$.

Therefore, from equation (9.1)
$$F_D = A_p C_D \rho u^2 / 2$$
$$= \frac{0.0625 \times 2 \times 1.2 \times 2^2}{2}$$
$$= \underline{0.30 \text{ N}}$$

## 9.1.2 Lift forces and lift coefficients

*Problem*

A circular kite of diameter 1.38 m and thickness 10 cm is held by a cord and flies at an angle to the horizontal. When the wind velocity is 30 km h$^{-1}$, and blows horizontally, the angle of the cord to the horizon is 60°, and the kite soars at right angles to the cord. If the mass of the kite is 510 g, and the tension in the cord holding the kite is 26.2 N, calculate the drag and lift forces on the kite, and the values of the drag and lift coefficients.

Density of air $= 1.2 \text{ kg m}^{-3}$

Viscosity of air $= 1.85 \times 10^{-5} \text{ kg m}^{-1} \text{s}^{-1}$

Acceleration due to gravity $= 9.81 \text{ m s}^{-2}$

## Solution

As the wind blows horizontally the drag force on the kite will also be horizontal and the lift force will be vertical. Five forces are acting on the kite, namely the drag force, $F_D$, the lift force, $F_L$, the force due to buoyancy, $F_B$, the tension in the cord, $F_T$, and the weight of the kite, $F_g$. At equilibrium these forces must balance. $F_T$ can therefore be resolved into horizontal and vertical components, as illustrated in Figure 9.1.

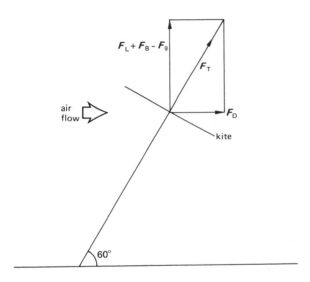

**Figure 9.1**

From a force balance in the horizontal direction, we have

$$F_D = F_T \cos 60°$$
$$= 26.2 \times 0.50 \text{ N}$$

and thus the drag force, $\quad F_D = 13.1 \text{ N}$

Now, $F_D$, can be related to the drag coefficient, $C_D$, by the equation (ref. 1, 2, 3).

$$\frac{F_D}{A_p} = \frac{C_D \rho u^2}{2}$$

whence $\quad C_D = 2F_D/(A_p \rho u^2) \quad\quad\quad (9.5)$

170  *Problems in fluid flow*

Now $A_p = A \cos 60°$ where $A$ is the surface area of one side of the kite.
Hence
$$A = \pi(1.38/2)^2 \text{ m}^2$$
$$= 1.50 \text{ m}^2$$
so that
$$A_p = 1.50 \times \cos 60° \text{ m}^2$$
$$= 0.75 \text{ m}^2$$
Further,
$$\rho = 1.2 \text{ kg m}^{-3}, \text{ and}$$
$$u = 30 \times 10^3 \text{ m h}^{-1}$$
$$= 8.33 \text{ m s}^{-1}$$

Substitution of the known terms into equation (9.5) gives
$$C_D = \frac{2 \times 13.1}{0.75 \times 1.2 \times 8.33^2}$$
i.e. drag coefficient, $\underline{C_D = 0.42}$

The weight of the kite,
$$F_g = mg$$
$$= 0.510 \times 9.81 \text{ N}$$
$$= 5.00 \text{ N}$$

The buoyancy force,
$$F_B = \rho \pi r^2 T g$$
$$= 1.2 \times \pi \times (1.38/2)^2 \times 0.1 \times 9.81 \text{ N}$$
$$= 1.76 \text{ N}$$

A balance of vertical force components, as shown in Figure 9.1, results in the equation
$$F_B + F_L - F_g = F_T \sin 60°$$
or
$$F_L = F_g - F_B + F_T \sin 60°$$

Substitution of the previously calculated values for $F_g$, $F_B$ and the given value for $F_T$ gives a lift force,
$$F_L = 5.00 - 1.76 + 26.2 \sin 60° \text{ N}$$
$$= \underline{25.9 \text{ N}}$$

$F_L$, can be related to the lift coefficient, $C_L$, by the equation (ref. 3)
$$\frac{F_L}{A_p} = \frac{C_L \rho u^2}{2}$$
or
$$C_L = 2 F_L / A_p \rho u^2 \qquad (9.6)$$
In this case
$$A_p = 1.5 \sin 60° \text{ m}^2$$
$$= 1.30 \text{ m}^2$$

Substitution of this and other known values into equation (9.6) enables the lift coefficient, $C_L$, to be determined.

Thus
$$C_L = \frac{2 \times 25.9}{1.30 \times 1.2 \times 8.33^2}$$
$$= \underline{0.48}$$

The kite thus experiences a drag force of 13.1 N and a lift force of 25.9 N. The experimental values of the drag and lift coefficients are 0.42 and 0.48 respectively.

### 9.1.3 The calculation of particle diameters from terminal settling velocities

*Problem*

A glass sphere of density 2280 kg m$^{-3}$ when falling through a solution of 98% glycerol reaches a terminal settling velocity of 0.040 m s$^{-1}$. Determine the diameter of the sphere.

Density of 98% glycerol = 1200 kg m$^{-3}$
Viscosity of 98% glycerol = 1.45 kg m$^{-1}$ s$^{-1}$
Acceleration due to gravity = 9.81 m s$^{-2}$

*Solution*

It can be shown from Newton's law that the terminal settling velocity, $U_t$, for the fall of a spherical particle of diameter $d$ and density $\rho_s$ through a fluid of density $\rho$ and viscosity $\mu$ is given by (ref. 10 and 11)

$$U_t^2 = \frac{4(\rho_s - \rho)gd}{3\rho C_D} \qquad (9.7)$$

where the drag coefficient, $C_D$, is a function of the Reynolds number, $Re$, which is defined as

$$Re = \frac{dU_t\rho}{\mu} \qquad (9.8)$$

Rearrangement of equation (9.7) gives

$$d = \frac{3\rho C_D U_t^2}{4(\rho_s - \rho)g} \qquad (9.9)$$

For the present problem, equation (9.9) contains two unknowns, $d$ and $C_D$. There are several procedures which may be used to overcome this. The first of these is a 'trial and error' method in which an approximate value for the particle diameter is assumed, the drag coefficient, $C_D$, calculated, and then substituted

## 172 Problems in fluid flow

into equation (9.9) to give the required diameter. Alternatively, equation (9.7) and the relationship $C_D = f(Re)$, which is available in graphical form (ref. 4, 5, 6) can be solved simultaneously in the following way.

The elimination of $d$ from equations (9.7) and (9.8) gives

$$U_t^2 = \frac{4(\rho_s - \rho)g}{3\rho C_D} \frac{Re\mu}{U_t \rho}$$

whence
$$C_D = \frac{4(\rho_s - \rho)g}{3\rho^2 U_t^3} Re\mu \qquad (9.10)$$

and
$$\log C_D = \log(Re) + \log\left(\frac{4(\rho_s - \rho)g\mu}{3\rho^2 U_t^3}\right) \qquad (9.11)$$

On a graph of $\log C_D$ versus $\log(Re)$, equation (9.11) is the equation of a straight line of unit slope passing through the point where $Re = 1$ and $C_D = [4(\rho_s - \rho)g\mu]/(3\rho^2 U_t^3)$.

The diameter $d$ does not occur in equation (9.11) but it can thus be determined by plotting equation (9.11) on a $\log C_D$ versus $\log Re$ graph (ref. 4, 5 and 6). The intersection of the plot of equation (9.11) with the curve for spheres gives the terminal Reynolds number from which $d$ can be calculated (ref. 12).

Another method (ref. 13) depends on the fact that

$$\frac{C_D}{2Re} = \frac{2\mu g}{3\rho^2 U_t^3}(\rho_s - \rho) \qquad (9.12)$$

(this equation follows by elimination of diameter $d$ from equations (9.7) and (9.8)). Equation (9.12) is independent of diameter. Log $Re$ is given as a function of $\log[C_D/(2Re)]$ (ref. 13) and the functions are plotted (ref. 14). The diameter of a sphere of known settling velocity can be determined by calculating $C_D/2Re$ using equation (9.12) and then finding the corresponding value of $Re$ from either table (ref. 13) or diagram (ref. 14). From $Re$, the diameter can be calculated.

For the present problem, using equation (9.12),

$$\frac{C_D}{2Re} = \frac{2 \times 1.45 \times 9.81(2280 - 1200)}{3 \times 1200^2 \times 0.04^3} = 111.1$$

From the plot of $C_D/2Re$ versus $Re$ (ref. 14) we have

$$Re = 0.32$$

Now
$$Re = dU_t\rho/\mu$$

whence
$$d = Re \cdot \mu/U_t\rho$$

$$= (0.32 \times 1.45)/(0.04 \times 1200) \text{ m}$$

$$= \underline{9.7 \times 10^{-3} \text{ m}}$$

The glass sphere therefore has a diameter of 9.7 mm.

## 9.1.4 The prediction of the terminal settling velocity of a spherical particle

*Problem*

Calculate the terminal settling velocity for a polyethylene sphere with a diameter of 3.4 mm when it falls freely through air.

Density of air $= 1.218 \text{ kg m}^{-3}$
Viscosity of air $= 1.73 \times 10^{-5} \text{ kg m}^{-1} \text{ s}^{-1}$
Density of polyethylene $= 910 \text{ kg m}^{-3}$
Acceleration due to gravity $= 9.81 \text{ m s}^{-2}$

*Solution*

From Newton's law the relationship between the diameter, $d$, of a particle and its terminal settling velocity, $U_t$, is given by the equation

$$U_t = \left(\frac{4(\rho_s - \rho)gd}{3\rho C_D}\right)^{\frac{1}{2}} \quad (9.13)$$

The relationship between drag coefficient, $C_D$, and $U_t$ may be expressed as

$$C_D = f\left(\frac{U_t \rho d}{\mu}\right) \quad (9.14)$$

In terms of the present problem, equations (9.13) and (9.14) have two unknowns, $U_t$ and $C_D$. They may therefore be solved simultaneously to determine the terminal settling velocity. Note however, that equation (9.14) is available in graphical form (ref. 3, 4 and 5).

An alternative procedure is to use the dimensionless group obtained by eliminating $U_t$ from equation (9.13) in the following manner. From equation (9.13)

$$C_D = \frac{4d(\rho_s - \rho)g}{(3 U_t^2 \rho)}.$$

Multiplication of both sides of this equation by

$$Re^2 = \left(\frac{U_t d \rho}{\mu}\right)^2$$

gives

$$C_D Re^2 = \frac{4 d^3 (\rho_s - \rho) \rho g}{3 \mu^2} \quad (9.15)$$

The right-hand side of this equation can be calculated for any particle settling in a fluid, so that the value of $C_D Re^2$ is known. Then $Re$ is deduced from a graph relating $C_D Re^2$ to $Re$ (ref. 14 and 15).

For the polyethylene particle settling through air, use of equation (9.15) gives

$$C_D Re^2 = \frac{4(3.4 \times 10^{-3})^3 (910 - 1.22) \times 1.22 \times 9.81}{3(1.73 \times 10^{-5})^2}$$

$$= 1.90 \times 10^6$$

From a graph of $C_D Re^2$ versus $Re$, it is found that

$$Re = 2200$$
$$= U_t \rho d / \mu.$$

Thus
$$U_t = Re\, \mu / \rho\, d$$
$$= (2200 \times 1.75 \times 10^{-5})/(1.22 \times 3.4 \times 10^{-3})$$
$$= 9.28 \text{ m s}^{-1}$$

The terminal settling velocity of the polyethylene sphere in air is therefore $9.3 \text{ m s}^{-1}$.

## 9.1.5 The effect of shape on the drag force experienced by non-spherical particles. Sphericity calculations

*Problem*

(i) Calculate the sphericity of a Raschig ring which is 2.0 cm long and has an internal diameter of 1.2 cm and an external diameter of 2.0 cm.

(ii) Four glass spheres, each having a diameter of 3.0 mm, are placed on a flat surface so that each sphere just makes contact with two others. The four spheres are glued at points of contact. Another sphere of the same size is placed on top of the four spheres and glued. The five spheres are then inverted, and a sixth sphere, identical to the others, is glued to the underside.

If the agglomerate is now allowed to settle through an oil of density $825 \text{ kg m}^{-3}$ and viscosity $1.21 \text{ kg m}^{-1} \text{s}^{-1}$, estimate the drag force experienced by the agglomerate when its settling velocity is $0.040 \text{ m s}^{-1}$.

Assume that when the agglomerate moves through the liquid, a plane formed by four coplanar spheres is perpendicular to the direction of movement.

*Solution*

(i) The most commonly used of the shape factors is the sphericity, $\psi$, (ref. 8 and 9) which for any particle is defined as the ratio of the surface area of a sphere which has a volume equal to that of the particle, to the surface area of the particle.

Thus
$$\psi = A_0/A$$
$$= \pi d_0^2/A$$

where $A$ is the surface area of the particle and $A_0$ is the surface area of the equivalent sphere. The diameter of the equivalent sphere is $d_0$.

If the particle has a volume $V$, then the volume of the equivalent sphere is also $V$, so that
$$V = \pi d_0^3/6$$
whence
$$d_0^2 = (6V/\pi)^{\frac{2}{3}}$$
and so
$$A_0 = \pi d_0^2$$
$$= \pi (6V/\pi)^{\frac{2}{3}}$$
and
$$\psi = \pi (6V/\pi)^{\frac{2}{3}}/A \tag{9.16}$$

The surface area of the Raschig ring equals the sum of the outside surface area, the inner surface area, and the surface area of the two annular ends.

Thus
$$A = 2\pi r_o L + 2\pi r_i L + 2(\pi r_o^2 - \pi r_i^2)$$
$$= 2\pi L(r_o + r_i) + 2\pi (r_o^2 - r_i^2)$$
$$= 2\pi (0.02)(0.01 + 0.006) + 2\pi (0.01^2 - 0.006^2) \text{ m}^2$$
$$= 2.41 \times 10^{-3} \text{ m}^2$$

Also, the volume of the Raschig ring, $V$, is given by
$$V = \pi r_o^2 L - \pi r_i^2 L$$
$$= \pi L(r_o^2 - r_i^2)$$
$$= \pi (0.02)(0.01^2 - 0.006^2) \text{ m}^3$$
$$= 4.02 \times 10^{-6} \text{ m}^3$$

Substituting the calculated values of the volume and surface area of the Raschig ring into equation (9.16) gives
$$\psi = \frac{\pi (6 \times 4.02 \times 10^{-6}/\pi)^{\frac{2}{3}}}{2.41 \times 10^{-3}}$$
$$= \underline{0.51}$$

The Raschig ring therefore has a sphericity of 0.51.

(ii) The method we shall use involves the calculation of the Reynolds number for the particle moving in the fluid, and the calculation of its sphericity. From these two values the drag coefficient, $C_D$, for the particle can be obtained from plots of Reynolds number versus drag coefficient, (ref. 4, 5 and 6). The drag force, $F_D$, is then calculated using the equation
$$F_D = A_p C_D u^2 \rho/2$$

To determine the Reynolds number, a mean projected diameter of the particle, $d_p$, is required. This may be defined as the diameter of a circle having the same area as the particle projects on a plane which is perpendicular to the direction of movement.

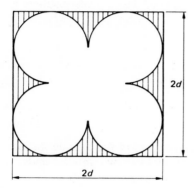

**Figure 9.2**

Figure 9.2 shows a projection of the agglomerate perpendicular to the direction of movement. If the diameter of each sphere is denoted by $d$, then the required projected area equals the area of a square with sides of length $2\,d$, minus the shaded area in the figure.

Now the area of the shaded region $= 4(3/4)(d^2 - \pi d^2/4)$
$$= 3(d^2 - \pi d^2/4)$$

The projected area, $A_p$, is thus given by

$$A_p = (2d)^2 - 3(d^2 - \pi d^2/4)$$
$$= d^2(1 + 3\pi/4)$$
$$= (3.0 \times 10^{-3})^2(1 + 3\pi/4)\ \text{m}^2$$
$$= 3.021 \times 10^{-5}\ \text{m}^2$$

Now $\quad A_p = \pi(d_p/2)^2$

So $\quad d_p = (4A_p/\pi)^{0.5}$
$$= (4 \times 3.021 \times 10^{-5}/\pi)^{0.5}\ \text{m}$$
$$= 6.20 \times 10^{-3}\ \text{m}$$

Thus the Reynolds number, $Re$, is given by

$$Re = d_p u \rho/\mu$$
$$= 6.20 \times 10^{-3} \times 0.04 \times 825/1.21$$
$$= 0.169$$

To determine the drag coefficient, $C_D$, the sphericity, $\psi$, for the agglomerate is also required. The surface area, $A$, of the agglomerate is given by

$$A = 6\pi d^2$$

and the volume of the agglomerate, $V$, by

$$V = 6(\pi d^3/6)$$
$$= \pi d^3$$

It follows from equation (9.16) that the sphericity, $\psi$, is given by

$$\psi = \pi (6\pi d^3/\pi)^{2/3}/6\pi d^2$$
$$= 6^{-1/3}$$
$$= 0.55$$

Using this value of the sphericity, and the Reynolds number previously determined, the drag coefficient, $C_D$, can be found from a plot of drag coefficient versus Reynolds number (ref. 4, 5 and 6). Thus $C_D = 105$.

Now the drag force, $F_D$, which the agglomerate experiences is given by

$$F_D = A_p C_D u^2 \rho/2$$
$$= 3.02 \times 10^{-5} \times 105 \times 0.04^2 \times 825/2 \text{ N}$$
$$= \underline{2.09 \times 10^{-3} \text{ N}}$$

## 9.1.6 Estimation of the settling velocity of particles under conditions of hindered settling

*Problem*

Glass spheres of 45 μm diameter are settling through water. If the suspension contains 60 kg of particles per 100 kg of water, estimate the maximum velocity reached by the particles assuming that hindered settling occurs.

Density of the glass = 2280 kg m$^{-3}$
Density of the water = 998 kg m$^{-3}$
Viscosity of the water = $1.25 \times 10^{-3}$ kg m$^{-1}$ s$^{-1}$
Acceleration due to gravity = 9.81 m s$^{-2}$

*Solution*

Several methods are described in the literature which may be applied to solve this problem. In the absence of specific information which can be applied to this system, the results of Steinour (ref. 16) will be used. It is first necessary to determine the range in which the motion of the particles lies i.e. the Stokes law range, when $500 < Re < 200\,000$, or the intermediate range, $2 < Re < 500$.

A convenient criterion for hindered settling in the Stokes law range is (ref. 17)

$$K = d\left[\frac{g\rho(\rho_s-\rho)\psi'^2}{\mu^2}\right]^{1/3} \qquad (9.17)$$

where $g$ is the acceleration of a particle from an external force, $\rho$ is the density of the slurry, $\rho_s$ is the particle density, $\mu$ is the liquid viscosity and $\psi'$ a viscosity correction factor for hindered settling. $\psi'$ is a function of the volume fraction of the liquid in the slurry, and empirical equations (and charts) relating $\psi'$ and the volume fraction ($e$) are available in various texts.

If $K$ is less than 3.3, the settling is in the Stokes law range.

For the present problem, we have $d = 45 \times 10^{-6}$ m, $g = 9.81$ m s$^{-2}$ and $\rho_s = 2280$ kg m$^{-3}$

To determine the density of the slurry, $\rho$, it is noted that the volume of 100 kg of water = $100/998$ m$^3$, and the volume of 60 kg of glass particles = $60/2280$ m$^3$.

The density of the glass plus water mixture therefore

$$= \frac{100+60}{(100/998)+(60/2280)} \text{ kg m}^{-3}$$

$$= 1265 \text{ kg m}^{-3}$$

Furthermore, the volume fraction of liquid in the slurry (or the voidage),

$$e = \frac{\text{volume of water}}{\text{volume of water} + \text{volume of glass}}$$

$$= \frac{100/998}{(100/998)+(60/2280)}$$

$$= 0.792$$

From charts of $e$ versus $\psi'$, or alternatively using the equation (ref. 17)

$$\psi' = \exp[-4.19(1-e)]$$

we have $\qquad \psi' = \exp[-4.19(1-0.79)]$

$$= 0.415$$

Substitution of known values into equation (9.17) gives

$$K = 45 \times 10^{-6}\left(\frac{9.81 \times 1265 \times (2280-1265) \times 0.415^2}{(1.25 \times 10^{-3})^2}\right)^{\frac{1}{3}}$$

$$= 0.50$$

The settling therefore occurs in the Stokes law range.

Using the viscosity correction factor, $\psi'$, for hindered settling, the following equation can now be used to estimate the terminal velocity of the particles relative to stationary coordinates:

$$U_H = \frac{gd^2 e\psi'(\rho_s - \rho)}{18\mu}$$

$$= \frac{9.81 \times (45 \times 10^{-6})^2 \times 0.79 \times 0.415 \times (2280 - 1265)}{18 \times 1.25 \times 10^{-3}} \text{ m s}^{-1}$$

$$= 2.94 \times 10^{-4} \text{ m s}^{-1}$$

An alternative method of estimating the settling velocity involves a correction factor, $R$, which incorporates both viscosity and density effects for a given slurry, and thus $R$ can be used to obtain a value for the settling velocity during hindered settling from the settling velocity calculated for unhindered, or free settling (ref. 18).

Thus in the present case, since the settling is in the Stokes law range, the terminal velocity for unhindered settling is given by

$$U_t = \frac{(\rho_s - \rho)gd^2}{18\mu}$$

and the settling velocity for *hindered* settling is given by

$$U_H = R\frac{(\rho_s - \rho)gd^2}{18\mu} \tag{9.18}$$

where $\rho$ and $\mu$ refer to the density and viscosity of the liquid, and not the slurry.

The correction factor $R$ may be obtained from the literature (ref. 16 and 18). In the present case, as $e = 0.79$, $R = 0.26$ (approximately). Substitution of known values into equation (9.18) then gives, for hindered settling

$$U_H = 0.26 \frac{(2280 - 998)(9.81)(45 \times 10^{-6})^2}{18 \times 1.25 \times 10^{-3}} \text{ m s}^{-1}$$

$$= 2.94 \times 10^{-4} \text{ m s}^{-1}$$

Note that these two methods are in fact identical and thus give exactly the same answer. The correction factor, $R$, can be calculated from the volume fraction of liquid, $e$.

Thus
$$R = e^2 \psi'$$
$$= e^2 \exp[-4.19(1-e)]$$

Other methods for estimating the terminal velocity of particles under conditions of hindered settling are also useful in some cases. For example Scholl (ref. 19) has related the hindered settling velocity of particles in non-flocculated suspensions ($U_H$) to the settling velocity in dilute suspensions ($U_t$) by the equation

$$U_H = e^n U_t$$

In this equation $e$ is again the volume fraction of liquid, and according to Scholl, (ref. 19), the exponent $n = 3.65$ for $e > 0.6$.

For the present problem, we therefore have

$$U_H = e^{3.65} U_t$$

$$= \frac{e^{3.65} g d^2 (\rho_s - \rho)}{18 \mu}$$

$$= \frac{0.79^{3.65} \times 9.81 \times (45 \times 10^{-6})^2 (2280 - 998)}{18 \times 1.25 \times 10^{-3}} \text{ m s}^{-1}$$

$$= \underline{4.79 \times 10^{-4} \text{ m s}^{-1}}$$

The terminal settling velocity under hindered settling conditions is approximately $3 \times 10^{-4}$ m s$^{-1}$, the actual value determined by calculation depending on the physical model, and therefore, the equation used.

### 9.1.7 The acceleration of a particle settling in a gravitational field

*Problem*

When a glass sphere of density 2280 kg m$^{-3}$ settles under gravity in 98% glycerol, it reaches a terminal velocity of 0.026 m s$^{-1}$. If the diameter of the sphere is $8.0 \times 10^{-3}$ m, how long does it take for it to reach 80% of its terminal settling velocity if it starts from zero velocity at the surface of the glycerol.

Density of 98% glycerol = 1200 kg m$^{-3}$
Viscosity of 98% glycerol = 1.45 kg m$^{-1}$ s$^{-1}$
Acceleration due to gravity = 9.81 m s$^{-2}$

*Solution*

When a body is settling in a liquid, the resultant force on the body in the vertical direction equals its weight minus the buoyant force of the displaced fluid and the drag force acting on the body. Since the resultant force is equal to the body's mass multiplied by its acceleration in the vertical direction, we can write

$$m \frac{du}{dt} = mg - V\rho g - F_D$$

where $m$ is the mass of the particle and $V$ its volume.

$$\text{Now } F_D = 0.5 A_p C_D \rho u^2 \qquad \text{(ref. 1, 2 and 3)}$$

Substituting for $F_D$ gives

$$m \frac{du}{dt} = mg - V\rho g - 0.5 A_p C_D \rho u^2 \qquad (9.19)$$

The various terms in equation (9.19) can be evaluated as follows:

$$m = (\pi/6)d^3 \rho_s$$
$$= (\pi/6)(8.0 \times 10^{-3})^3(2280) = 6.112 \times 10^{-4} \text{ kg}$$
$$V = (\pi/6)d^3$$
$$= (\pi/6)(8.0 \times 10^{-3})^3 = 2.681 \times 10^{-7} \text{ m}^3$$

Now as the terminal velocity of the particle is 0.026 m s$^{-1}$, when the particle has reached 80% of this its velocity is 0.0208 m s$^{-1}$. The Reynolds number corresponding to this is given by

$$Re = du\rho/\mu$$
$$= (8.0 \times 10^{-3})(0.0208)(1200)/1.45 = 0.14$$

Thus the drag coefficient for all velocities up to 0.0208 m s$^{-1}$ (and beyond) can be calculated from the equation

$$C_D = 24/Re$$
$$= 24\mu/d\rho u$$
$$= 24 \times 1.45/[(8.0 \times 10^{-3})1200 \times u]$$
$$= 3.625/u$$

Furthermore
$$A_p = \pi(d/2)^2$$
$$= \pi(8.0 \times 10^{-3}/2)^2 \text{ m}^2$$
$$= 5.027 \times 10^{-5} \text{ m}^2$$

These values can now be substituted into equation (9.19) to give

$$(6.112 \times 10^{-4})\frac{du}{dt} = (6.112 \times 10^{-4})(9.81)$$
$$- (2.681 \times 10^{-7})(1200)(9.81)$$
$$- 0.5(5.027 \times 10^{-5})(3.625)(1200)u$$

or
$$\frac{du}{dt} = 4.646 - 178.9\, u$$

To determine the time elapsed for the particle to reach 80% of its terminal velocity, i.e. 0.0208 m s$^{-1}$, the above equation must be integrated. Rearrangement gives

$$dt = \frac{du}{4.646 - 178.9u}$$

Integration gives

$$t = \int_0^{0.0208} \frac{du}{4.646 - 178.9u}$$

$$= -\frac{1}{178.9} [\log_e (4.646 - 178.9\, u)]_0^{0.0208} \text{ s}$$

$$= 0.0090 \text{ s}$$

Therefore it will take 9 milliseconds for the particle to reach 80% of its terminal settling velocity.

## 9.2 Student Exercises

1  Spherical particles of a material which had a density of 1996 kg m$^{-3}$ were separated into various sizes and their terminal settling velocities in water were determined. The following results were obtained:

| Particle diameter (m) | Terminal settling velocity (m s$^{-1}$) |
|---|---|
| 1.0 × 10$^{-6}$ | 5.51 × 10$^{-7}$ |
| 1.0 × 10$^{-5}$ | 5.50 × 10$^{-5}$ |
| 1.0 × 10$^{-4}$ | 5.01 × 10$^{-3}$ |
| 0.5 × 10$^{-3}$ | 0.06 |
| 1.0 × 10$^{-3}$ | 0.15 |
| 3.33 × 10$^{-3}$ | 0.33 |
| 8.12 × 10$^{-3}$ | 0.54 |
| 15 × 10$^{-3}$ | 0.71 |

Use these results to prepare a graph, on logarithmic coordinates, showing the relationship between Reynolds number and drag coefficient, $C_D$. Compare your graph with those available (ref. 4, 5 and 6).

Density of water = 998 kg m$^{-3}$
Viscosity of water = 1.01 × 10$^{-3}$ kg m$^{-1}$ s$^{-1}$
Acceleration due to gravity = 9.81 m s$^{-2}$

2  A balloon which has a diameter of 1.80 metres and a mass of 0.740 kg is held by a thin cord. If the cord makes an angle of 16° to the horizontal, and the wind blows horizontally, estimate the wind velocity.

Density of air = 1.22 kg m$^{-3}$
Viscosity of air = 1.75 × 10$^{-5}$ kg m$^{-1}$ s$^{-1}$
Acceleration due to gravity = 9.81 m s$^{-2}$

3  A sample of sand is composed of spherical particles which range in diameter from $5.0 \times 10^{-6}$ m to $60.0 \times 10^{-6}$ m. Prepare a graph which relates the terminal settling velocity in air to the diameter of the particles.

Density of sand sample = 3100 kg m$^{-3}$
Density of air = 1.22 kg m$^{-3}$
Viscosity of air = 1.82 × 10$^{-5}$ kg m$^{-1}$ s$^{-1}$
Acceleration due to gravity = 9.81 m s$^{-2}$

4  A falling ball viscometer consists of an all glass vessel which is 8 cm in diameter and 30 cm deep. It is filled with a liquid whose viscosity is required, kept at a constant temperature, and the time taken for a 'ball' to fall a known distance down the centre of the column of test liquid is recorded.

In one test on an organic liquid, which had a density of 1260 kg m$^{-3}$, a sapphire sphere which was 3.175 mm in diameter passed between the graduation marks, which were 15.0 cm apart, in 15.05 seconds. Determine the viscosity of the liquid.

Density of sapphire = 3980 kg m$^{-3}$
Acceleration due to gravity = 9.81 m s$^{-2}$

5  An aqueous slurry containing 7.5% by weight of calcium carbonate is dispersed thoroughly and then allowed to settle. The height of the line of demarcation between clear liquid and opaque slurry is measured as a function of time, and the following results obtained.

| Time (minutes) | Height of interface (metres) |
|---|---|
| 0 | 0.31 |
| 14 | 0.28 |
| 28 | 0.25 |
| 47 | 0.21 |
| 62 | 0.18 |
| 72 | 0.16 |

If the particles of calcium carbonate are assumed to be spherical and of uniform size, use Steinour's equation (ref. 16) to estimate their diameter.

184  *Problems in fluid flow*

6  A glass sphere which has a diameter of 1.5 mm settles in a hydrocarbon which has a density of 960 kg m$^{-3}$ and a viscosity of 0.093 kg m$^{-1}$ s$^{-1}$. The sphere starts from rest.
(i) What is its velocity when it has reached the end of the Stokes law range?
(ii) How long does it take to reach the end of the Stokes law range?
(iii) How far has it fallen when it reaches the end of the Stokes law range?

Density of glass = 2280 kg m$^{-3}$
Acceleration due to gravity = 9.81 m s$^{-2}$

7  A polyethylene sphere of 3.4 mm diameter is thrown through air at an initial velocity of 5 m s$^{-1}$ in a weightless environment. How far will it have travelled in 1 minute?

Density of air = 1.22 kg m$^{-3}$
Density of polyethylene = 910 kg m$^{-3}$
Viscosity of air = 1.82 × 10$^{-5}$ kg m$^{-1}$ s$^{-1}$


## 9.3  References

References are abbreviated as follows:

McC, S & H = W. L. McCabe, J. C. Smith and P. Harriott, Unit Operations of Chemical Engineering, 4th edition, McGraw-Hill, 1985

D, G and S = J. F. Douglas, J. M. Gasiorek and J. A. Swaffield, Fluid Mechanics, Pitman, 1983

F et al. = A. S. Foust, L. A. Wenzel, C. W. Clump, L. Maus and L. B. Andersen, Principles of Unit Operations, 2nd edition, John Wiley and Sons, 1980

C and R (2) = J. M. Coulson and J. F. Richardson, Chemical Engineering, Vol. 2, 3rd edition, Pergamon Press, 1977

1. McC, S & H, page 129
2. F et al., page 611
3. D, G and S, page 309
4. McC, S & H, page 131
5. F et al., page 612
6. D, G and S, page 319
7. C and R (2), page 89
8. F et al., page 711
9. McC, S & H, page 137
10. McC, S & H, page 142
11. F et al., page 613
12. F et al., page 614
13. C and R (2), page 99
14. C and R (2), page 98
15. D, G and S, page 326
16. H. H. Steinour, *Industrial and Engineering Chemistry*, 1944, **36**, 618, 840, 901
17. W. L. McCabe and J. C. Smith, Unit Operations of Chemical Engineering, 3rd edition, McGraw-Hill, 1976, page 157
18. F et al., page 615
19. K. H. Scholl, *Chem. Ing. Techn.*, 1976, **48**, 149–50

## 9.4 Notation

| Symbol | Description | Unit |
|---|---|---|
| $A$ | surface area of particle or body | $m^2$ |
| $A_p$ | projected area of particle in plane perpendicular to fluid flow direction | $m^2$ |
| $A_0$ | surface area of sphere which has same volume as particle | $m^2$ |
| $C_D$ | drag coefficient | — |
| $C_L$ | lift coefficient | — |
| $d$ | diameter of particle | m |
| $d_0$ | diameter of sphere which has same volume as particle | m |
| $d_p$ | mean projected diameter of particle | m |
| $e$ | void fraction, or volume fraction of liquid in slurry | — |
| $F_B$ | buoyancy force acting on particle or body | N |
| $F_D$ | drag force acting on particle or body | N |
| $F_g$ | weight of particle or body | N |
| $F_L$ | lift force acting on body | N |
| $F_T$ | tension in cord (see Figure 9.1) | N |
| $g$ | acceleration due to gravity | $m\,s^{-2}$ |
| $K$ | criterion for hindered settling | — |
| $L$ | length of one side of a cube, or characteristic dimension of a particle | m |
| $m$ | mass of body or particle | kg |
| $n$ | exponent in Scholls equation (ref. 19) | — |
| $R$ | viscosity and density correction factor for hindered settling | — |
| $r$ | radius of spherical particle | m |
| $r_i$ | inside radius | m |
| $r_o$ | outside radius | m |
| $Re$ | Reynolds number | — |
| $T$ | thickness of kite | m |
| $u$ | mean linear velocity of particle or fluid | $m\,s^{-1}$ |
| $U_H$ | settling velocity of particle in hindered settling | $m\,s^{-1}$ |
| $U_t$ | terminal settling velocity of particle | $m\,s^{-1}$ |
| $V$ | volume of particle | $m^3$ |
| $\rho$ | density of fluid | $kg\,m^{-3}$ |
| $\rho_s$ | density of solid, or particle | $kg\,m^{-3}$ |
| $\mu$ | viscosity of fluid | $kg\,m^{-1}\,s^{-1}$ |
| $\psi$ | sphericity | — |
| $\psi'$ | viscosity correction factor | — |

# 10 Sedimentation and classification

## 10.1 Worked Examples

### 10.1.1 The determination of settling velocities from a single batch sedimentation test

*Problem*

A slurry of calcium carbonate containing 50 g of dry powder in 1 dm$^3$ of water was thoroughly mixed and then allowed to settle. The height of the interface between the clear liquid and the suspended solids was measured at various times after settling began, and the results are listed in Table 10.1. Use these results to prepare a curve showing the relationship between the settling rate and the concentration of calcium carbonate in a slurry.

**Table 10.1**

| Time (hours) | Height of interface (m) |
|---|---|
| 0 | 1.10 |
| 0.5 | 1.03 |
| 1.0 | 0.96 |
| 2.0 | 0.82 |
| 3.0 | 0.68 |
| 4.0 | 0.54 |
| 5.0 | 0.42 |
| 6.0 | 0.35 |
| 7.0 | 0.31 |
| 8.0 | 0.28 |
| 9.0 | 0.27 |
| 10.0 | 0.27 |

*Solution*

From the experimental data given the height of the interface ($Z$) is plotted as a function of time ($t$) as shown in Figure 10.1.

*Sedimentation and classification* 187

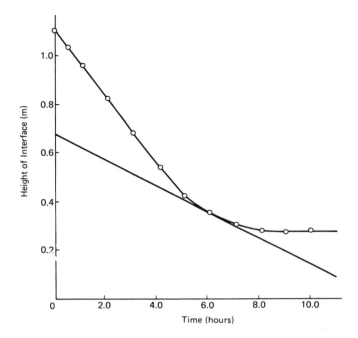

**Figure 10.1**

Now from the Kynch theory of settling, (ref. 1 and 2) we have that

$$C_L Z_i = C_0 Z_0 \qquad (10.1)$$

where $C_0$ and $Z_0$ are the initial concentration and height of the suspended solids in the batch-settling test and $C_L$ is the concentration in a layer of the suspension which, according to the Kynch theory, is limiting the rate of flow of solids. $Z_i$ is obtained, for any value of time $t = t_L$, from the plot given in Figure 10.1, since the tangent to the curve at any point $(t_L, Z_L)$ intersects the $Z$ axis at $Z = Z_i$.

From the data provided,

$$C_0 = 50 \text{ g dm}^{-3}$$
$$= 50 \text{ kg m}^{-3}$$

and $\quad Z_0 = 1.10 \text{ m}$

Thus $\quad C_0 Z_0 = 50 \times 1.10 \text{ kg m}^{-2}$

$$= 55.0 \text{ kg m}^{-2}$$

Rearrangement of equation (10.1) gives

$$C_L = C_0 Z_0 / Z_i$$

188  Problems in fluid flow

so that
$$C_L = 55.0/Z_i \text{ kg m}^{-3} \qquad (10.2)$$

For several points on the curve given in Figure 10.1, tangents are drawn and both the slope and the intercept are measured. One such tangent, through the point ($t_L = 6$ hours, $Z = 0.35$ m) is shown on the figure.

From the Kynch theory, the settling velocity, $U_c$, calculated as the slope of the tangent, is the settling velocity corresponding to a concentration $C_L$, where $C_L$ is calculated from the intercept $Z_i$, using equation (10.2).

The values of the settling velocities and concentrations determined from tangents drawn through seven points on the curve in Figure 10.1 are shown in Table 10.2.

**Table 10.2**

| Time (hours) | $Z_i$ (m) | Settling velocity (m h$^{-1}$) | $C_L$ (kg m$^{-3}$) |
|---|---|---|---|
| 1.5 | 1.10 | 0.139 | 50.0 |
| 3.0 | 1.10 | 0.139 | 50.0 |
| 5.0 | 0.99 | 0.113 | 55.6 |
| 5.5 | 0.84 | 0.085 | 65.5 |
| 6.0 | 0.64 | 0.048 | 85.9 |
| 7.0 | 0.53 | 0.031 | 103.8 |
| 8.0 | 0.43 | 0.019 | 127.9 |

A plot of settling velocity versus the slurry concentration is given in Figure 10.2, as is required.

### 10.1.2 The estimation of the minimum area required for a continuous thickener in order to effect a given rate of separation of solids from a suspension

*Problem*

A continuous thickener is required to concentrate a slurry of calcium carbonate in water from a solids content of 50 kg m$^{-3}$ to 130 kg m$^{-3}$, and to produce a clear overflow containing no calcium carbonate. The density of the dry calcium carbonate is 2300 kg m$^{-3}$, and a single batch sedimentation test produced the experimental results listed in Table 10.1.

If the thickener is fed slurry at a rate of 0.06 m$^3$ s$^{-1}$, determine the minimum thickener area that is required, and the flow rate of clarified water.

Density of water = 998 kg m$^{-3}$

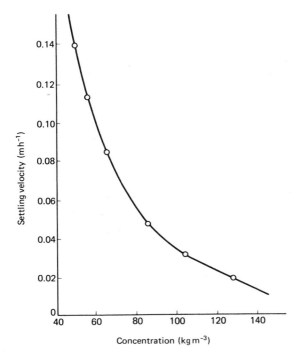

**Figure 10.2**

*Solution*

In a continuously operating thickener, solids must settle through layers of all concentrations between that of the feed and that of the final underflow. Settling rates decrease with increasing concentration, and therefore decrease as depth increases in a continuous thickener. Furthermore, the mass of liquid displaced upwards will vary throughout a thickener due to changes in the amount of liquid removed from layer to layer and the volume fraction of liquid in each layer. Thickener design depends on determining the concentration of the layer having the lowest capacity for the settling of solids. Sufficient cross-sectional area must be provided so that the settling rate exceeds the linear liquid upflow rate in this layer.

If $U$ is the mass ratio of liquid to solid at any level in a continuous thickener, and $V$ the mass ratio of liquid to solid in the underflow, the mass of liquid displaced upwards, per kg of solid in the feed, $= U - V$.

If the solids flow rate in the feed $= F$ kg s$^{-1}$, the total upflow rate of liquid $= F(U - V)$ kg s$^{-1}$ and the volumetric upflow rate is given by

$$Q = (F/\rho)(U - V)$$

where $\rho$ is the density of the liquid.

The linear upflow rate $= Q/A$
$$= (F/A\rho)(U - V)$$
where $A$ is the cross-sectional area of the thickener.

Now the linear upflow rate must be less than the settling rate at all levels,

i.e. $\quad\quad\quad\quad\quad\quad\quad (F/\rho A)(U - V) < U_t$

or $\quad\quad\quad\quad\quad\quad\quad A > F(U - V)/(\rho U_t)\quad\quad\quad\quad (10.3)$

Now $U$, $V$ and $U_t$ vary from level to level within the thickener. The problem is therefore to determine the maximum value of the right hand side of equation (10.3) between $U = $ feed concentration and $V = $ underflow concentration, remembering that $U_t$ is a function of $U$.

The concentration units given in Table 10.2 can be converted to a mass ratio of liquid/solids as follows.

For a concentration of slurry $= C$ kg m$^{-3}$, the volume of solids per m$^3$ of slurry $= C/\rho_s$.

The volume of liquid per m$^3$ of slurry thus $= 1 - C/\rho_s$.

The mass of liquid $= \rho(1 - C/\rho_s)$

and finally mass of liquid/mass of solids $= \rho(1 - C/\rho_s)/C$
$$= \rho(1/C - 1/\rho_s).$$

Thus as the concentration of solids in the underflow $= 130$ kg m$^{-3}$,

$$V = 998\,(1/130 - 1/2300)\text{ kg liquid/kg solids}$$
$$= 7.24 \text{ kg liquid/kg solids}$$

Concentrations read from Figure 10.2 can likewise be converted to a (mass of liquid)/(mass of solid) basis, as given in Table 10.3. Also included in this table are the corresponding settling velocities, which are also read off Figure 10.2.

**Table 10.3**

| $C_L$ (kg m$^{-3}$) | $U$ | Settling velocity (m h$^{-1}$) |
|---|---|---|
| 50.0 | 19.53 | 0.139 |
| 55.0 | 17.71 | 0.115 |
| 60.0 | 16.20 | 0.098 |
| 65.0 | 14.92 | 0.083 |
| 70.0 | 13.82 | 0.071 |
| 75.0 | 12.87 | 0.062 |
| 80.0 | 12.04 | 0.055 |
| 85.0 | 11.31 | 0.049 |
| 90.0 | 10.65 | 0.043 |
| 100.0 | 9.55 | 0.034 |

Now the solids flow rate in the feed,

$$F = 0.06 \times 50 = 3.0 \text{ kg s}^{-1}$$
$$= 10\,800 \text{ kg h}^{-1}$$

The right hand side of equation (10.3) can therefore be calculated for values of $U$ between the feed concentration, ($C = 50$ kg m$^{-3}$, $U = 19.53$) and the underflow concentration ($C = 130$ kg m$^{-3}$, $V = 7.24$) in order to determine its maximum value. These calculations are listed in Table 10.4, for the values of $U$ given in Table 10.3.

To determine the minimum value of the thickener area required, the values of $F(U - V)/\rho U_t$ can be plotted against $U_t$, as shown in Figure 10.3, to show that the maximum value of $A$ occurs at approximately <u>1005 m$^2$</u>.

Thus in order to concentrate the calcium carbonate slurry from a concentration of 50 kg dm$^{-3}$ to 130 kg dm$^{-3}$, the area of the thickener should be at least 1005 m$^2$.

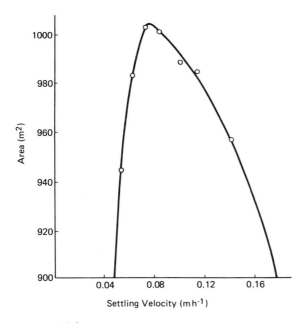

**Figure 10.3**

To determine the flow rate of the clarified water, a flow-rate balance gives

$$Q_F = Q_u + Q_o$$

Now the input flow rate = 0.06 m$^3$ s$^{-1}$, and the solids content of this is

**Table 10.4**

| $C_L$ (kg m$^{-3}$) | $U_t$ (m h$^{-1}$) | $U$ | $U-V$ | $A$ (m²) |
|---|---|---|---|---|
| 50 | 0.139 | 19.53 | 12.29 | 956.8 |
| 55 | 0.115 | 17.71 | 10.47 | 985.2 |
| 60 | 0.098 | 16.20 | 8.96 | 989.4 |
| 65 | 0.083 | 14.92 | 7.68 | 1001.3 |
| 70 | 0.071 | 13.82 | 6.58 | 1002.9 |
| 75 | 0.062 | 12.87 | 5.63 | 982.7 |
| 80 | 0.055 | 12.04 | 4.80 | 944.4 |
| 85 | 0.049 | 11.31 | 4.07 | 898.9 |
| 90 | 0.043 | 10.65 | 3.41 | 858.1 |
| 100 | 0.034 | 9.55 | 2.31 | 735.2 |

50 kg m$^{-3}$. The flow rate of calcium carbonate in the feed thus $= 50 \times 0.06 = 3.0$ kg s$^{-1}$

The solids content of the effluent is 130 kg m$^{-3}$.

The flow rate of the effluent therefore $= 3.0/130 = 0.0231$ m³ s$^{-1}$

The volumetric flow rate of the overflow therefore $= 0.06 - 0.0231 = 0.037$ m³ s$^{-1}$

The volumetric flow rate of the clarified water which leaves the thickener as overflow is therefore 0.037 m³ s$^{-1}$.

## 10.1.3 Settling velocities and the classification of materials which have different settling velocities

*Problem*

An intimate mixture of two different ores of densities 2600 kg m$^{-3}$ and 5100 kg m$^{-3}$ is finely ground and given a screen analysis. The various fractions collected on the screens are then analysed chemically to determine the amount of each ore in each size fraction, and in this way the following particle size analysis for each component of the original mixture is obtained.

Ore of density = 2600 kg m$^{-3}$

| Particle diameter, (μm) | 15 | 25 | 35 | 45 | 55 | 65 | 75 | 85 | 95 |
|---|---|---|---|---|---|---|---|---|---|
| Weight percent of particles undersize | 0.00 | 22.0 | 35.0 | 47.0 | 59.0 | 68.0 | 75.0 | 81.0 | 100.0 |

Ore of density = 5100 kg m$^{-3}$

| Particle diameter, (μm) | 25 | 35 | 45 | 55 | 65 | 75 | 85 | 95 |
|---|---|---|---|---|---|---|---|---|
| Weight percent of particles undersize | 0.00 | 21.0 | 33.5 | 46.0 | 57.5 | 67.0 | 75.0 | 100.0 |

The ratio of the heavy ore to the light ore in the mixture was by analysis, 1:5 by weight.

(i) If the ores are to be separated from the original mixture by elutriation using water at a velocity of $4.00 \times 10^{-3}$ m s$^{-1}$ and a temperature of 18 °C, what are the particle size range and composition of the bottom product, and what fraction of the heavy ore in the original ore mixture is contained in this bottom product?

(ii) If the ores are to be separated by elutriation using a liquid of density 1500 kg m$^{-3}$ and viscosity $6.25 \times 10^{-4}$ kg m$^{-1}$ s$^{-1}$ what liquid velocity is required to produce the maximum amount of bottom product which contains none of the lighter ore particles, what is the size range of this bottom product, and what fraction of the heavy ore in the original ore mixture is contained in this bottom product?

(iii) In the separation described in part (ii) above, what is the mass fraction of light ore in the overhead product?

(iv) If a complete separation of the two components in the original ore mixture is to be achieved using an unhindered settling operation in which the particles settle at their terminal velocities, what is the minimum density of the fluid that would be required? Assume that the viscosity of this fluid will be greater than that of water.

Density of water at 18 °C = 998.6 kg m$^{-3}$
Viscosity of water at 18 °C = $1.03 \times 10^{-3}$ kg m$^{-1}$ s$^{-1}$
Acceleration due to gravity = 9.81 m s$^{-2}$

*Solution*

(i) The sizes of the particles of each ore which have a settling velocity equal to the upward velocity of the water, i.e. $4.00 \times 10^{-3}$ m s$^{-1}$, have to be determined. This calculation in turn depends on the Reynolds number ($Re$) for the system.

Now for the largest particles, $d = 9.5 \times 10^{-5}$ m. The maximum Reynolds number will therefore be given by

$$Re = dU_t\rho/\mu$$
$$= (9.5 \times 10^{-5})(0.004)(998.6)/(1.03 \times 10^{-3})$$
$$= 0.37$$

Thus the relative flow of fluid and particles will always be in the Stokes law range, i.e. where $Re < 2$.

It can be shown from Newton's Law that the terminal settling velocity, $U_t$, for the fall of a spherical particle of diameter $d$ and density $\rho_s$ through a fluid of density $\rho$ and viscosity $\mu$ is given by

$$U_t = \sqrt{\frac{4(\rho_s - \rho)gd}{3\rho C_D}} \quad (10.4)$$

where the drag coefficient, $C_D$, is a function of the Reynolds number, $Re$. Furthermore, in the Stokes law range the drag coefficient is related to the Reynolds number as follows:

$$C_D = 24/Re$$

Substitution for $C_D$ in equation (10.4) gives

$$U_t = \frac{gd^2(\rho_s - \rho)}{18\mu}$$

i.e.
$$d = \sqrt{\frac{18\mu U_t}{g(\rho_s - \rho)}} \quad (10.5)$$

For the heavy particles, it follows that the smallest particles which would settle through water flowing at $4.00 \times 10^{-3}$ m s$^{-1}$ have a diameter,

$$d = \left(\frac{18 \times 1.03 \times 10^{-3} \times 0.004}{9.81 \times (5100 - 998.6)}\right)^{\frac{1}{2}} \text{m}$$

$$= 4.29 \times 10^{-5} \text{ m}$$

$$= \underline{42.9 \ \mu\text{m}}$$

For the lighter particles, in exactly the same way it follows that their smallest diameter,

$$d = \left(\frac{18 \times 1.03 \times 10^{-3} \times 0.004}{9.81 \times (2600 - 998.6)}\right)^{\frac{1}{2}} \text{m}$$

$$= 6.87 \times 10^{-5} \text{ m}$$

$$= \underline{68.7 \ \mu\text{m}}$$

When the water flows at a velocity of $4.00 \times 10^{-3}$ m s$^{-1}$ the particles of the heavy ore with a diameter less than 42.9 $\mu$m, and the particles of the light ore less than 68.7 $\mu$m in diameter, will be carried away in the water stream. So, the overall size range of particles in the bottom product = 42.9 to 95.0 $\mu$m.

If the particle size distribution data for the two ores are plotted, interpolation of the plots shows that 30% of the heavy ore has a diameter less than 42.9 $\mu$m and will be carried away, and 71.5% of the light ore has a diameter less than 68.7 $\mu$m and so will also be carried away by the water stream.

From the problem statement it is seen that the ratio of heavy ore to light ore in the original mixture is equal to 1:5, so that the composition of the original mixture is

heavy ore, $\dfrac{1}{1+5} \times 100 = 16.67\%$

light ore, $\dfrac{5}{1+5} \times 100 = 83.33\%$

Thus the heavy ore removed by the water per 100 kg of the original mixture $= 0.30 \times 16.67$ kg $= 5.00$ kg, whereas the light ore removed by the water per 100 kg of the original mixture $= 0.715 \times 83.33$ kg $= 59.58$ kg.

It follows that the amount of heavy ore not removed by the water stream

$$= 16.67 - 5.00 \text{ kg per 100 kg of original mixture}$$
$$= 11.67 \text{ kg per 100 kg of original mixture}$$

The amount of light ore not removed by the water stream

$$= 83.33 - 59.58 \text{ kg per 100 kg}$$
$$= 23.75 \text{ kg per 100 kg}$$

Hence for the bottom product remaining after the elutriation, the composition by weight is

heavy ore, $\dfrac{11.67}{11.67+23.75} \times 100 = 32.95\%$

light ore, $\dfrac{23.75}{11.67+23.75} \times 100 = 67.05\%$

The original ore mixture contained 16.67 kg of heavy ore per 100 kg. As 11.67 kg of this remains in the bottom product, the fraction of the heavy ore in the original mixture which remains is $11.67/16.67 = 0.70$

To summarise the particle size range of the bottom product is 42.9 $-95.0$ $\mu$m, the composition of the bottom product is; heavy ore, 33.0%, light ore, 67.0%, and 70% of the heavy ore in the original ore mixture is retained in the bottom product.

(ii) The heavy ore will settle faster, and therefore the terminal settling velocity of the largest particles of the light ore in the mixture will be the factor which determines the liquid velocity required to produce the maximum amount of bottom product containing only heavy ore. The largest particles of light ore have a diameter of $9.5 \times 10^{-5}$ m, and to determine the settling velocity of these the table of log $Re$ as a function of log $[Re^2 (R^1/\rho u^2)]$, for spherical particles (ref. 3 and 4) can again be used, as was the case for worked example 9.1.4.

Thus using reference 3, we have

$$Re^2(R^1/\rho u^2) = 2d^3\rho g(\rho_s - \rho)/(3\mu^2)$$
$$= 2(9.5 \times 10^{-5})^3(1500)(2600 - 1500)$$
$$\times (9.81)/3(0.625 \times 10^{-3})^2$$
$$= 23.68$$

Now $\log 23.68 = 1.3745$
From Table 3.2 of ref. 3 we find that

$$\log Re = 0.2136$$

whence
$$Re = 1.6352$$

Now
$$Re = dU_t\rho/\mu$$

so terminal settling velocity of particles,

$$U_t = Re\,\mu/d\rho$$
$$= (1.6352 \times 0.625 \times 10^{-3})/(9.5 \times 10^{-5} \times 1500)\,\text{m s}^{-1}$$
$$= 0.0072 \text{ m s}^{-1}$$
$$= 7.2 \text{ mm s}^{-1}$$

This is also the required liquid velocity.

To calculate the minimum size of the bottom product, equation 10.5 above can be used as the Reynolds number in this case is clearly much less than 2.

Thus in this case

$$d = [18\mu U_t/g(\rho_s - \rho)]^{\frac{1}{2}}$$
$$= [18(0.625 \times 10^{-3})(0.0072)/(9.81)(2600 - 1500)]^{\frac{1}{2}} \text{ m}$$
$$= 8.66 \times 10^{-5} \text{ m}$$
$$= 86.6 \text{ }\mu\text{m}$$

The size range is therefore only from 86.6 µm to 95 µm. The interpolation shows that 75% of the heavy ore has a diameter less than 86.6 µm. Thus 25% of the heavy ore in the original mixture is contained in the bottom product.
(iii) The overhead product contains all of the light ore in the original mixture, and 75% of the heavy ore in the original mixture.

It has been shown in part (i) of this problem that the original ore mixture contains 16.67 kg of heavy ore and 83.33 kg of light ore per 100 kg.

Thus the mass fraction of the light ore in the overhead product

$$= \frac{83.33}{83.33 + (0.75 \times 16.67)}$$
$$= 0.87$$

(iv) Complete separation of the two components in the original ore mixture can only be achieved by a classification process if the settling velocity of the smallest particles of the heavy ore is greater than the settling velocity of the largest particle of light ore. The limiting situation for complete separation is therefore when these two settling velocities are equal.

If one denotes the heavy ore by 'a' and the light ore by 'b', then since the terminal velocity of any spherical particle is given by

$$U_t = \sqrt{\frac{4(\rho_s - \rho)gd}{3\rho C_D}}$$

it follows that for the limiting situation,

$$\sqrt{\frac{4(\rho_a - \rho)gD_a}{3\rho C_{Da}}} = \sqrt{\frac{4(\rho_b - \rho)gD_b}{3\rho C_{Db}}}$$

or

$$\frac{D_a}{D_b} = \frac{(\rho_b - \rho)C_{Da}}{(\rho_a - \rho)C_{Db}} \tag{10.6}$$

where $\rho_a$ and $\rho_b$ are the densities of a and b and $\rho_a > \rho_b$, $D_a$ is the diameter of the smallest 'a' particles, $D_b$ is the diameter of the largest 'b' particles, and $C_{Da}$ and $C_{Db}$ are the drag coefficients of the smallest 'a' particles and largest 'b' particles respectively.

Since the drag coefficient is essentially constant at high Reynolds numbers, it follows that for particles settling at high Reynolds numbers, equation (10.6) reduces to

$$\frac{D_a}{D_b} = \frac{\rho_b - \rho}{\rho_a - \rho} \tag{10.7}$$

However if it is assumed that settling is in the Stokes law range (i.e. $Re < 2$), then

$$C_D = \frac{24}{Re} \tag{10.8}$$

and it therefore follows that in this flow regime

$$C_{Da} = \frac{24\mu}{D_a U_t \rho} \tag{10.9}$$

and

$$C_{Db} = \frac{24\mu}{D_b U_t \rho} \tag{10.10}$$

(Note that the terminal velocities $(U_t)$ of the smallest 'a' and largest 'b' particles are equal in the limiting situation).

Substitution of equations (10.9) and (10.10) into equation (10.6) gives, for the Stokes law range,

$$\frac{D_a}{D_b} = \frac{(\rho_b - \rho)D_b}{(\rho_a - \rho)D_a}$$

or

$$\left(\frac{D_a}{D_b}\right)^2 = \frac{\rho_b - \rho}{\rho_a - \rho} \qquad (10.11)$$

As noted above, complete separation of the two types of particles will only occur if the settling velocity of the smallest particles of the heavy ore, 'a', is greater than the settling velocity of the largest particles of the light ore, 'b'. When this situation exists

$$\left(\frac{D_a}{D_b}\right)^2 > \frac{\rho_b - \rho}{\rho_a - \rho} \qquad (10.12)$$

For the present problem,

$$D_a = 2.5 \times 10^{-5}\,\text{m}, \quad D_b = 9.5 \times 10^{-5}\,\text{m},$$
$$\rho_a = 5100\,\text{kg m}^{-3}, \quad \rho_b = 2600\,\text{kg m}^{-3}$$

Substitution into equation (10.12) gives

$$\left(\frac{2.5 \times 10^{-5}}{9.5 \times 10^{-5}}\right)^2 > \frac{2600 - \rho}{5100 - \rho},$$

from which it follows that $\rho > 2414\,\text{kg m}^{-3}$

Therefore if a complete separation of the two components in the original ore mixture is to be achieved by classification, the fluid must have a density greater than $2414\,\text{kg m}^{-3}$.

To complete the solution, the assumptions that settling is in the Stokes law range should be checked using the method given in problem 9.1.4 to determine $C_D Re^2$ and thence $Re$.

## 10.1.4 Density variations in a settling suspension

### Problem

A sample of quartz has a particle size distribution as follows:

| Mass % of sample with a diameter $> d$ | Diameter $d$ (μm) |
|---|---|
| 5.0 | 30.2 |
| 11.8 | 21.4 |
| 20.2 | 17.4 |
| 24.2 | 16.2 |
| 28.5 | 15.2 |
| 37.6 | 12.3 |
| 61.8 | 8.8 |

## Sedimentation and classification

A homogeneous slurry containing $40.0\,\text{kg m}^{-3}$ of this quartz sample in water is allowed to settle. If the particles are assumed to be spherical what is the density of the slurry 25.0 cm from the surface of the liquid after a settling time of 18.5 minutes?

Density of water = $998\,\text{kg m}^{-3}$
Density of quartz = $2660\,\text{kg m}^{-3}$
Viscosity of water = $1.01 \times 10^{-3}\,\text{kg m}^{-1}\,\text{s}^{-1}$
Acceleration due to gravity = $9.81\,\text{m s}^{-2}$

### Solution

If $\rho_t$ is the density of the slurry at the depth $h$ after a settling time $t$, $\rho_s$ is the density of the solid particles, and $\rho$ the fluid density, then it may be shown (ref. 5) that

$$w = \rho_s(\rho_t - \rho)/(\rho_s - \rho) \quad (10.13)$$

where $w$ is the mass of particulate matter per m³ of the suspension, at the given depth, which has a diameter, $d$, corresponding to the settling time $t$. In addition, if $w_0$ is the mass of the particulate matter per m³ of suspension at zero time, when the suspension is homogeneous, then

$$\phi = w/w_0 \quad (10.14)$$

where $\phi$ is the mass fraction of the original sample of particles which has a diameter less than $d$.
Combining equations (10.13) and (10.14) gives

$$\phi = \rho_s(\rho_t - \rho)/(\rho_s - \rho)w_0 \quad (10.15)$$

Now $\phi$ can be calculated from the settling time and the original particle size distribution. Thus for a settling time of 18.5 minutes, or 1110 seconds, the terminal settling velocity, $U_t$, is given by

$$U_t = h/t$$
$$= 25.0 \times 10^{-2}/1110 = 2.25 \times 10^{-4}\,\text{m s}^{-1}$$

Assuming that the settling is in the Stokes law range, which may be checked later, the diameter of the particles of quartz with a terminal settling velocity of $U_t$ is given by

$$d = [18\,\mu U_t/g(\rho_s - \rho)]^{\frac{1}{2}} \quad (10.16)$$
$$= [18(1.01 \times 10^{-3})(2.25 \times 10^{-4})/9.81\,(2660 - 998)]^{1/2}\,\text{m}$$
$$= 15.8 \times 10^{-6}\,\text{m}$$

From the particle size distribution data given an interpolation shows that

about 25.9% of the quartz has a diameter greater than 15.8 μm, so that
$$\phi = 1 - 0.259 = 0.741$$
The substitution of known values into equation (10.15) gives
$$0.741 = 2660\,(\rho_t - 998)/(2660 - 998)\,40.0$$
whence $\quad\quad\quad\quad \rho_t = 1017\,\text{kg m}^{-3}$

To check that the settling of particles with a diameter of $15.8 \times 10^{-6}$ m is within the Stokes law range, we have
$$\begin{aligned} Re &= dU_t\rho/\mu \\ &= (15.8 \times 10^{-6})(2.25 \times 10^{-4})(998)/(1.01 \times 10^{-3}) \\ &= 3.5 \times 10^{-3} \end{aligned}$$

Thus the settling is in fact in the range for which equation (10.16) is applicable. Therefore the density of the suspension at a depth of 25.0 cm after a settling time of 18.5 minutes is $\underline{1017\,\text{kg m}^{-3}}$.

## 10.1.5 The determination of particle size distribution using a sedimentation method

*Problem*

A sample of cement powder is completely dispersed in dry ethanol containing 0.2% of calcium chloride as dispersant. The suspension is transferred to a cylindrical vessel, thoroughly agitated, and then allowed to settle in a constant temperature room held at 20 °C.

The tip of a pipette is lowered into the cement suspension to a depth of 20 cm below the surface and at various times a 10 cm³ sample of the suspension at this depth is removed without disturbing the settling operation. The distance between the tip of the pipette and the surface of the liquid remains unchanged during the removal of samples because of the large diameter of the cylindrical vessel.

The liquid from the removed samples is evaporated, and the mass of solids (cement plus calcium chloride) which remains is determined. Altogether seven samples are removed and evaporated, and the results are listed in Table 10.5. In this table zero time is the time at which agitation ceases and settling begins.
(i) Determine the particle size distribution of the cement sample.
(ii) Show graphically that these particles have a logarithmic normal size distribution.

Density of cement = $3100\,\text{kg m}^{-3}$
Density of dispersant solution = $790\,\text{kg m}^{-3}$
Viscosity of dispersant solution = $1.2 \times 10^{-3}\,\text{kg m}^{-1}\,\text{s}^{-1}$

## Table 10.5

| Sample number | Settling time (seconds) | Mass of solids removed (g) |
|---|---|---|
| 0 | 0 | 0.1911 |
| 1 | 45 | 0.1586 |
| 2 | 135 | 0.1388 |
| 3 | 495 | 0.1109 |
| 4 | 1 875 | 0.0805 |
| 5 | 6 900 | 0.0568 |
| 6 | 66 600 | 0.0372 |
| 7 | 86 400 | 0.0359 |

*Solution*

It can be shown from Newton's Law that the terminal velocity ($U_t$) for the fall of a spherical particle of diameter $d$ and density $\rho_s$ through a fluid of density $\rho$ and viscosity $\mu$ is given by

$$U_t = [4(\rho_s - \rho)gd/3\rho\, C_D]^{\frac{1}{2}} \tag{10.17}$$

where the drag coefficient, $C_D$, is a function of the Reynolds number, $Re$, which is defined as

$$Re = d\rho U_t/\mu$$

Now in the Stokes law range, where $Re$ is less than 2, the drag coefficient is related to the Reynolds number as follows:

$$C_D = 24/Re$$

Substitution for $C_D$ in equation (10.17) gives

$$U_t = gd^2(\rho_s - \rho)/18\mu \tag{10.18}$$

When a settling particle falls a distance $h$ in time $t$, then

$$U_t = h/t \tag{10.19}$$

We can combine equations (10.18) and (10.19) to give

$$d = \left[\frac{18\mu h}{gt(\rho_s - \rho)}\right]^{\frac{1}{2}} \tag{10.20}$$

This equation is only valid for Reynolds numbers less than 2 i.e. the Stokes law range.

If we start with a uniform dispersion of cement in a liquid, and after time $t$ a small sample of the suspension is removed at depth $h$, the sample obtained will be an average sample for all particles of diameter less than $d$, where $d$ is given by equation (10.20).

Thus for sample number one of Table 10.5,

$$d = \left[\frac{18 \times 1.2 \times 10^{-3} \times 20 \times 10^{-2}}{9.81 \times 45 \times (3100 - 790)}\right]^{\frac{1}{2}} m$$

$$= 6.51 \times 10^{-5} \, m$$

$$= 65.1 \, \mu m$$

For the seven samples of Table 10.5, diameters shown in Table 10.6 were calculated in this way.

**Table 10.6**

| Sample number | Settling time(s) | $d$ (μm) |
|---|---|---|
| 1 | 45 | 65.1 |
| 2 | 135 | 37.6 |
| 3 | 495 | 19.6 |
| 4 | 1875 | 10.1 |
| 5 | 6900 | 5.3 |
| 6 | 66 600 | 1.7 |
| 7 | 86 400 | 1.5 |

The mass of solids obtained from the evaporation of ethanol from each sample contains both calcium chloride and cement. In 1000 cm³ of alcohol there is 2.0 g of calcium chloride, so that each sample of 10 cm³ removed by the pipette contains 0.0200 g of calcium chloride. The mass of cement in each dry solids sample is obtained by subtracting this amount of calcium chloride from the total solids listed in Table 10.5. The mass of cement in each sample is thus given in Table 10.7.

**Table 10.7** Cement content of sedimentation samples

| Sample number | Mass of cement (g) |
|---|---|
| 0 | 0.1711 |
| 1 | 0.1386 |
| 2 | 0.1188 |
| 3 | 0.0909 |
| 4 | 0.0605 |
| 5 | 0.0368 |
| 6 | 0.0172 |
| 7 | 0.0159 |

Sample 0 represents the mass of the cement in the original, completely dispersed, suspension. The ratio of the mass of cement in each other sample to the mass of cement in sample 0 therefore gives the mass per cent of cement particles of less than diameter $d$ in each sample. When combined with the calculations of $d$ given in Table 10.6 the particle size distribution as shown in Table 10.8 is obtained.

Figure 10.4 gives the distribution plotted on probability coordinates. The straight line shows that there is a logarithmic normal distribution of particle sizes.

**Table 10.8**

| $d$ (μm) | Mass percent of particles of diameter less than $d$ |
|---|---|
| 65.1 | 81.0 |
| 37.6 | 69.4 |
| 19.6 | 53.1 |
| 10.1 | 35.4 |
| 5.3 | 21.5 |
| 1.7 | 10.1 |
| 1.5 | 9.3 |

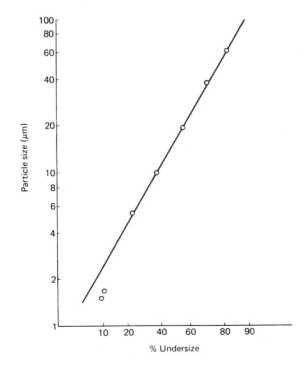

**Figure 10.4**

A basic assumption of the solution procedure used for this problem is that the settling must occur in the Stokes law range. This is confirmed by calculating the Reynolds number for the larger particles. The settling velocity, $U_t$, for the particles of diameter $= 65.1\ \mu m$ is given by

$$U_t = h/t$$
$$= 20 \times 10^{-2}/45\ \text{m s}^{-1}$$
$$= 4.4 \times 10^{-3}\ \text{m s}^{-1}$$

Thus
$$Re = dU_t\rho/\mu$$
$$= (65.1 \times 10^{-6})(4.4 \times 10^{-3})(790)/(1.2 \times 10^{-3})$$
$$= 0.19$$

The Reynolds number for particles with a diameter less than 65.1 $\mu m$ must be less than 0.19. Since 81% of the particles have a diameter less than 65.1 $\mu m$, settling does occur substantially in the Stokes law range, which requires a Reynolds number less than 2.

## 10.1.6 The determination of the particle size distribution of a suspended solid by measurement of the mass rate of sedimentation. The sedimentation balance

*Problem*

A slurry containing an ore thoroughly dispersed in water was allowed to settle in a cylindrical vessel. A pan was located centrally in the vessel at a depth of 25.00 cm below the liquid surface, and was so placed that all the particles which settled would fall onto the pan. The ore particles which settled on the pan were weighed continuously. The results, which are presented in Table 10.9, give the

**Table 10.9**

| Settling time (s) | Total mass of ore settled (g) |
|---|---|
| 114 | 0.1429 |
| 150 | 0.2010 |
| 185 | 0.2500 |
| 276 | 0.3564 |
| 338 | 0.4208 |
| 396 | 0.4781 |
| 456 | 0.5354 |
| 582 | 0.6139 |
| 714 | 0.6563 |
| 960 | 0.7277 |

total mass of ore which settled in the pan after a given settling time. After 960 seconds, 0.0573 g of material remained in the suspension.

Determine the particle size distribution of the ore, expressed as a table of $W$ versus $d$, where $W$ is the mass % of ore sample which has a diameter greater than $d$. Assume that all the ore particles are spherical.

Density of water = 998 kg m$^{-3}$
Density of ore = 2398 kg m$^{-3}$
Viscosity of water = $1.01 \times 10^{-3}$ kg m$^{-1}$ s$^{-1}$
Acceleration due to gravity = 9.81 m s$^{-2}$

*Solution*

The principle of this method is the determination of the rate at which particles settle out of a homogeneous suspension. If $W$ is the mass percent of a sample which has a diameter greater than $d$, then the mass percent, $P$, which has settled out at time $t$, is the sum of *two* terms. The first term consists of the mass % of all particles which have a terminal settling velocity greater than the sedimentation depth divided by the settling time. The second term derives from those particles which have a terminal settling time which is less than the first group, but which have settled because they started off at some intermediate position, and not at the liquid surface. It is readily shown (ref. 6) that, if $P$ is the percentage of the particles that have settled by a time $t$, and if $W$ is defined as above, then

$$P = W + t\frac{dP}{dt}$$

or
$$W = P - t\frac{dP}{dt} \quad (10.21)$$

Now from the data given in Table 10.9, $P$ can be calculated. A plot of $P$ versus $t$, followed by graphical differentiation, will give $dP/dt$ at various values of $t$, and thus $W$ can be found at these times.

The particle size, $d$, can also be calculated from the $t$ values, using any method relating terminal settling velocity to diameter. In virtually all cases where this procedure is used the settling is in the Stokes law range, and the following equation can be used to estimate the particle size.

$$d = \sqrt{18\mu h/(\rho_s - \rho)gt} \quad (10.22)$$

where $h$ is the distance between balance pan and free surface of the slurry.

From the results listed in Table 10.9, we can calculate the diameters corresponding to the various settling times as follows:

$$\mu = 1.01 \times 10^{-3} \,\text{kg}\,\text{m}^{-1}\,\text{s}^{-1}$$
$$\rho_s = 2398 \,\text{kg}\,\text{m}^{-3}$$
$$\rho = 998 \,\text{kg}\,\text{m}^{-3}$$
$$h = 25.0 \times 10^{-2} \,\text{m}$$

Substitution into equation (10.22) gives the diameters listed in Table 10.10.

As it is stated that 0.0573 g of particles remained in suspension, the total mass of material in the original slurry = 0.7277 + 0.0573 or 0.7850 g. The mass percent of the sample ($P$) which has settled out at time $t$ is therefore readily calculated. For example, for a settling time of 114 s,

$$P = (0.1429/0.7850) \times 100\%$$
$$= 18.2\%$$

Other values of $P$, corresponding to the given settling times, are given in Table 10.10. The only other quantity required for equation (10.21) is $dP/dt$. This may be found by plotting $P$ versus $t$ and calculating the gradients at

**Table 10.10**

| Settling time (s) | Diameter (µm) | P % | $\dfrac{dP}{dt}$ (s$^{-1}$) |
|---|---|---|---|
| 114 | 53.9 | 18.2 | 0.160 |
| 150 | 47.0 | 25.5 | 0.160 |
| 185 | 42.3 | 31.8 | 0.160 |
| 276 | 34.6 | 45.5 | 0.141 |
| 338 | 31.3 | 53.6 | 0.129 |
| 396 | 28.9 | 60.9 | 0.121 |
| 456 | 26.9 | 68.2 | 0.099 |
| 582 | 23.8 | 78.2 | 0.066 |
| 714 | 21.5 | 83.6 | 0.038 |
| 960 | 18.6 | 92.7 | 0.025 |

various $t$ values. Such a plot is given in Figure 10.5, which shows one gradient only, namely that corresponding to $t = 456$ s. Other values of $dP/dt$ obtained in this way are given in Table 10.10.

Substitution of the now calculated values of $P$, $t$ and $dP/dt$ into equation (10.21) allows $W$, the mass percentage of ore sample which has a diameter greater than $d$, to be calculated. For example, for $t = 456$ s,

$$W = P - t\frac{dP}{dt}$$
$$= 68.2 - 456\,(0.099)\%$$
$$= 23.1\%$$

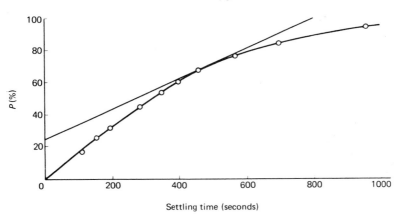

**Figure 10.5**

The required table of $W$ versus $d$, where $W$ is the mass % of ore sample of diameter greater than $d$, is given in Table 10.11.

Note that for the calculation of $W$ it is frequently preferable to plot $P$ versus $\log t$ and to determine the gradient $dP/d(\log t)$.

Equation (10.21) can be rearranged to give

$$W = P - \frac{dP}{d(\log t)}$$

in this case (ref. 6).

**Table 10.11**

| $W$ (mass %) | $d$ (µm) |
|---|---|
| 6.6 | 34.6 |
| 10.0 | 31.3 |
| 13.0 | 28.9 |
| 23.1 | 26.9 |
| 39.8 | 23.8 |
| 56.5 | 21.5 |
| 68.5 | 18.6 |

It was assumed that in this problem the settling was in the Stokes law range. This can now be checked.

$$Re = dU_t\rho/\mu$$

For the particles which settled in 114 s,

$$U_t = (25 \times 10^{-2})/114 \text{ m s}^{-1}$$
$$= 2.19 \times 10^{-3} \text{ m s}^{-1}$$

and $d = 53.9 \times 10^{-6}$ m

Therefore $Re = (53.9 \times 10^{-6})(2.19 \times 10^{-3})(998)/(1.01 \times 10^{-3})$
$= 0.117$

The settling is therefore in the required range for the application of equation (10.22) to all particles having a settling time of 114 s or more.

## 10.1.7 The decanting of homogeneous suspensions to obtain particles of a given size range

*Problem*

Dilute aqueous sodium oxalate forms a useful dispersing liquid for kaolin. A slurry of kaolin in this liquid is poured into a measuring cylinder to a depth of 40 cm, thoroughly mixed, and allowed to settle. After 600 seconds the supernatant liquid is decanted off until the level drops to one half of its initial value. Fresh dispersing liquid is added to bring the liquid level in the measuring cylinder back to 40 cm, and a second settling time of 600 seconds is allowed to pass before a second decanting is made. This process is repeated until the supernatant liquid is clear.

(i) What fraction of those particles in the original slurry which have a particle size of $15 \times 10^{-6}$ m will be removed in the first and second decantings?

(ii) After six decantings, what fraction of those particles in the original slurry having a particle size of $15 \times 10^{-6}$ m remain in the suspension?

Density of dispersant = 1002 kg m$^{-3}$
Density of kaolin = 2240 kg m$^{-3}$
Viscosity of dispersant = $1.01 \times 10^{-3}$ kg m$^{-1}$ s$^{-1}$
Acceleration due to gravity = 9.81 m s$^{-2}$

*Solution*

(i) Consider a cylindrical vessel containing a depth $h_1$ of the initial homogeneous suspension. When $h_2$ m of liquid is decanted no particle coarser than $d$ will be removed, so that

$$d = \sqrt{18\mu U_t/[g(\rho_s - \rho)]} \qquad (10.23)$$

where, if $t$ is the settling time before decanting,

$$U_t = h_2/t$$

However, particles of diameter less than $d$ (e.g. with a diameter equal to $xd$, where $x < 1$) will have fallen a distance $h$ in time $t$, where

$$xd = \sqrt{18\mu(h/t)/[g(\rho_s - \rho)]} \qquad (10.24)$$

From equations (10.23) and (10.24) it follows that (ref. 7)

$$x^2 = h/h_2$$

The fraction of particles of size $xd$ removed

$$= (h_2 - h)/h_1$$
$$= (h_2/h_1)(1 - x^2) \qquad (10.25)$$

and the fraction of particles of size $xd$ still left in suspension thus

$$= 1 - (h_2/h_1)(1 - x^2) \qquad (10.26)$$

When the suspension is made up to its original level, mixed, and a second fraction removed after a further time $t$, the fraction of particles of size $xd$ removed will be

$$= (h_2/h_1)(1 - x^2)[1 - (h_2/h_1)(1 - x^2)] \qquad (10.27)$$

Therefore the fraction of the original particles in the slurry, of diameter $xd$, removed in the first plus the second decanting is given by the sum of the equations (10.25) and (10.27).

For the present problem, $d$ can be found using equation (10.23)

Thus $\quad d^2 = 18(1.01 \times 10^{-3})(3.33 \times 10^{-4})(9.81(2240 - 1002))$ m$^2$

$\quad$ (as $\quad U_t = h_2/t = 20 \times 10^{-2}/600 = 3.33 \times 10^{-4}$ m s$^{-1}$)

Thus $\quad d = 22.33 \times 10^{-6}$ m

Therefore as $15.0 \times 10^{-6} = x(22.33 \times 10^{-6})$

$$x = 0.672$$

The fraction of particles of size $xd$ removed in the first decanting, from equation (10.25),

$$= (20/40)(1 - 0.672^2)$$
$$= 0.274$$

The fraction of particles of size $xd$ removed in the second decanting, from equation (10.27)

$$= (20/40)(1 - 0.672^2)[1 - (20/40)(1 - 0.672^2)]$$
$$= 0.199$$

Therefore the fraction of the original particles in the slurry having a particle

210  *Problems in fluid flow*

size of $15 \times 10^{-6}$ m which will be removed in the first plus the second decanting

$$= 0.274 + 0.199 = 0.473$$

or  $\underline{47.3\%}$  of the particles.

(ii) From equation (10.26), after one decanting, the fraction of particles of diameter $xd$ left in suspension

$$= 1 - (h_2/h_1)(1 - x^2)$$

After 2 decantings, the fraction of particles of diameter $xd$ left in suspension therefore

$$= [1 - (h_2/h_1)(1 - x^2)]^2$$

and so on until after 6 decantings, the fraction

$$= [1 - (h_2/h_1)(1 - x^2)]^6 \qquad (10.28)$$

Substitution of known values into equation (10.28) gives that the fraction remaining

$$= [1 - (20/40)(1 - 0.672^2)]^6$$
$$= 0.146$$

Thus, after six decantings, 14.6% of the particles in the original suspension which have a diameter of $15 \times 10^{-6}$ m will still remain in the suspension vessel.

These calculations assume that the kaolin particles are spherical. They also assume that the settling occurs in the Stokes law range. To check this, the Reynolds number, $Re$, is given by

$$Re = dU_t\rho/\mu$$
$$= 22.33 \times 10^{-6} (3.33 \times 10^{-4})(1002)/(1.01 \times 10^{-3})$$
$$= 7.4 \times 10^{-3}$$

Thus the settling is in the Stokes law range.

## 10.2 Student Exercises

1  An aqueous slurry containing 7.5% by weight of calcium carbonate is thoroughly dispersed and then allowed to settle. The height of the line of demarcation between clear liquid and opaque slurry is measured as a function of time and results, as shown in Table 10.12, were obtained.

Prepare a graph of sedimentation rate versus concentration, and thus determine the settling rate of the calcium carbonate at a concentration of 9.7 kg m$^{-3}$.

## Table 10.12

| Time (minutes) | Height of interface (m) |
|---|---|
| 0 | 0.31 |
| 14 | 0.28 |
| 28 | 0.25 |
| 47 | 0.21 |
| 62 | 0.18 |
| 72 | 0.16 |
| 84 | 0.14 |
| 98 | 0.12 |
| 118 | 0.10 |
| 132 | 0.09 |
| 150 | 0.08 |

Density of the calcium carbonate = $2710 \text{ kg m}^{-3}$
Density of water = $998 \text{ kg m}^{-3}$
Viscosity of water = $1.001 \times 10^{-3} \text{ kg m}^{-1} \text{s}^{-1}$
Acceleration due to gravity = $9.81 \text{ m s}^{-2}$

2  A sludge from a water treatment plant contains 6 kg of water per kg of solids. It is to be thickened to reduce the water content to 2 kg per kg of solids, and a thickener of area 70.0 m² is available for this. Calculate the maximum rate at which the sludge can be fed to this thickener if it operates continuously and produces solid-free overflow at all times.

Batch sedimentation tests on a laboratory scale gave the following results for the sludge.

## Table 10.13

| Concentration (kg water/kg solid) | Sedimentation rate (m h$^{-1}$) |
|---|---|
| 6.1 | 0.61 |
| 5.6 | 0.46 |
| 5.0 | 0.30 |
| 4.0 | 0.15 |
| 2.5 | 0.08 |

3  A batch sedimentation test on a slurry of an ore (65 g) in water (1 dm³) gave the following results:

**Table 10.14**

| Settling time (hours) | Height of interface (m) |
|---|---|
| 0.00 | 0.400 |
| 0.05 | 0.333 |
| 0.10 | 0.263 |
| 0.20 | 0.160 |
| 0.30 | 0.105 |
| 0.40 | 0.083 |
| 0.50 | 0.067 |
| 0.60 | 0.057 |
| 0.70 | 0.047 |
| 0.80 | 0.037 |
| 0.90 | 0.027 |
| 1.00 | 0.017 |
| 1.10 | 0.012 |
| 1.20 | 0.010 |

The density of the ore was 2500 kg m$^{-3}$. Determine the minimum diameter for a thickener which is to process 50 tonnes of solids per day in a feed of 65 g dm$^{-3}$, and which gives an underflow concentration of 400 g dm$^{-3}$ and a clear overflow.

Density of water = 998 kg m$^{-3}$

4  A homogeneous suspension containing 10.80 g of polyvinylchloride particles made up to one litre with isobutyl alcohol was allowed to settle in a measuring cylinder, and the pipette method used to determine the particle size distribution. Ten 10 cm$^3$ samples were removed from 20.0 cm below the surface of the suspension at various times, and the mass of polyvinylchloride in each sample was determined. The results are given in Table 10.15.

**Table 10.15**

| Sampling time (seconds) | Mass of solids in 10.0 cm$^3$ sample (g) |
|---|---|
| 800 | 0.0022 |
| 1153 | 0.0023 |
| 2492 | 0.0108 |
| 3201 | 0.0305 |
| 4431 | 0.0529 |
| 7203 | 0.0767 |
| 11 990 | 0.0961 |
| 26 130 | 0.1026 |
| 51 220 | 0.1053 |
| 320 100 | 0.1069 |

Plot the particle size distribution of the polyvinylchloride on linear coordinates.

Density of isobutyl alcohol = 801 kg m$^{-3}$
Viscosity of isobutyl alcohol = 4.703 × 10$^{-3}$ kg m$^{-1}$ s$^{-1}$
Density of polyvinylchloride = 1400 kg m$^{-3}$
Acceleration due to gravity = 9.81 m s$^{-2}$

5  Diamond powder is to be separated from talc using a double-cone classifier, with water as the separating fluid. If the diameter of the largest particles of diamond or talc is 60 × 10$^{-6}$ m, estimate the minimum diameter that the particles can have for a complete separation of the two materials.

Density of diamond = 3500 kg m$^{-3}$
Density of talc = 2405 kg m$^{-3}$
Density of water = 998 kg m$^{-3}$
Viscosity of water = 1.10 × 10$^{-3}$ kg m$^{-1}$ s$^{-1}$

6  A sample of silica has the particle size distribution shown in Table 10.16

**Table 10.16**

| $d$ (μm) | Mass % > $d$ |
|---|---|
| 18 | 100 |
| 20 | 99.3 |
| 24 | 84.4 |
| 30 | 69.6 |
| 40 | 48.9 |
| 50 | 34.1 |
| 60 | 25.2 |
| 70 | 17.8 |
| 80 | 13.3 |
| 90 | 10.4 |
| 100 | 8.1 |

If a homogeneous slurry of this silica in water is poured into a vessel which is 2.0 m deep and left undisturbed, what percentage of the silica will have settled out on the bottom of the vessel after 30 minutes?

Density of silica = 2650 kg m$^{-3}$
Density of water = 998 kg m$^{-3}$
Viscosity of water = 1.01 × 10$^{-3}$ kg m$^{-1}$ s$^{-1}$
Acceleration due to gravity = 9.81 m s$^{-2}$

## 214 Problems in fluid flow

7  A sample of silica containing particles of diameter $60 \times 10^{-6}$ m mixed with particles of diameter $80 \times 10^{-6}$ m is to be separated in a water elutriator. If the elutriator tube has a diameter of 5.0 cm, and the water flows at a superficial mean linear velocity of $0.32 \text{ cm s}^{-1}$, calculate
(i) the fraction of silica of particle size $60 \times 10^{-6}$ m which is removed in the water stream, and
(ii) the fraction of silica of particle size $80 \times 10^{-6}$ m which is removed in the water stream.

Density of silica = $2650 \text{ kg m}^{-3}$
Density of water = $998 \text{ kg m}^{-3}$
Viscosity of water = $1.01 \times 10^{-3} \text{ kg m}^{-1} \text{ s}^{-1}$
Acceleration due to gravity = $9.81 \text{ m s}^{-2}$

## 10.3 References

Two references are abbreviated as follows:

C and R (2) = J. M. Coulson and J. F. Richardson, Chemical Engineering, Vol. 2, 3rd edition, Pergamon, 1980

T. Allen = T. Allen, Particle Size Measurement, 2nd edition, Chapman and Hall, 1975

1. C and R (2), page 185
2. A. S. Foust, L. A. Wenzel, C. W. Clump, L. Maus and L. B. Andersen, Principles of Unit Operations, 2nd edition, John Wiley and Sons, 1980, page 633
3. C and R (2), page 97
4. J. F. Douglas, J. M. Gasiorek, and J. A. Swaffield, Fluid Mechanics, Pitman, 1980, page 326
5. T. Allen, page 192
6. T. Allen, page 222
7. T. Allen, page 236

## 10.4 Notation

| Symbol | Description | Unit |
|---|---|---|
| $A$ | cross-sectional area of vessel | $\text{m}^2$ |
| $C$ | concentration of a slurry | $\text{kg m}^{-3}$ |
| $C_0$ | initial concentration of a slurry | $\text{kg m}^{-3}$ |
| $C_L$ | concentration, in rate limiting layer, of a settling slurry | $\text{kg m}^{-3}$ |
| $C_D$ | drag coefficient | — |
| $C_{Da}$ | drag coefficient of the smallest 'a' particles | — |
| $C_{Db}$ | drag coefficient of the largest 'b' particles | — |
| $d$ | particle diameter | m |
| $D_a$ | diameter of smallest type 'a' particles | m |
| $D_b$ | diameter of largest type 'b' particles | m |
| $F$ | mass flow rate of feed stream | $\text{kg s}^{-1}$ |
| $g$ | acceleration due to gravity | $\text{m s}^{-2}$ |
| $h$ | depth | m |
| $h_1$ | initial depth of a suspension | m |

| Symbol | Description | Unit |
|---|---|---|
| $h_2$ | decrease in depth of suspension on decanting | m |
| $P$ | mass percentage of particles which have settled after time $t$ | — |
| $Q$ | volumetric upflow rate in a continuous thickener | $m^3 s^{-1}$ |
| $Q_F$ | volumetric feed flow rate | $m^3 s^{-1}$ |
| $Q_u$ | volumetric underflow rate | $m^3 s^{-1}$ |
| $Q_o$ | volumetric overflow rate | $m^3 s^{-1}$ |
| $R^1$ | resistance per unit projected area of particle, (ref. 3) | $N m^{-2}$ |
| $Re$ | Reynolds number | — |
| $t$ | settling time | s |
| $t_L$ | a particular value of settling time on Figure 10.1 | s |
| $U$ | mass ratio of liquids to solids at any level in a thickener | — |
| $U_c$ | settling velocity, corresponding to concentration $C_L$, in a thickener | $m s^{-1}$ |
| $U_t$ | terminal settling velocity | $m s^{-1}$ |
| $V$ | mass ratio of liquid to solid in underflow | — |
| $W$ | mass percentage of particles with diameter greater than $d$ | — |
| $w$ | mass of solids per unit volume of slurry | $kg m^{-3}$ |
| $w_0$ | initial mass of solids per unit volume of slurry | $kg m^{-3}$ |
| $x$ | ratio of a particle diameter less than $d$, to $d$ | — |
| $Z$ | height of clear liquid/slurry interface in sedimentation | m |
| $Z_i$ | height of rate limiting layer of a settling slurry | m |
| $Z_0$ | initial height of liquid in a batch settling test | m |
| $\mu$ | fluid viscosity | $kg m^{-1} s^{-1}$ |
| $\rho$ | fluid density | $kg m^{-3}$ |
| $\rho_a$ | density of type 'a' particles | $kg m^{-3}$ |
| $\rho_b$ | density of type 'b' particles | $kg m^{-3}$ |
| $\rho_s$ | density of solids or particles | $kg m^{-3}$ |
| $\rho_t$ | density of a slurry at depth $h$ after settling for time $t$ | $kg m^{-3}$ |
| $\phi$ | mass fraction of particles which have a diameter less than $d$ | — |

# 11 Fluidisation

## 11.1 Worked Examples

### 11.1.1 Particulate and aggregative fluidisation

*Problem*

Establish whether particulate or aggregative fluidisation occurs in the following systems:
(i) Catalyst particles of diameter $0.3 \times 10^{-3}$ m and density 2600 kg m$^{-3}$ are fluidised in a high pressure gas. The vessel used is cylindrical, with an inside diameter of 0.95 m. The height of the catalyst bed under minimum fluidising conditions is 3.25 m, and the superficial gas velocity under minimum fluidising conditions is $2.10 \times 10^{-2}$ m s$^{-1}$. Under these operating conditions the gas has a viscosity of $2.21 \times 10^{-5}$ kg m$^{-1}$ s$^{-1}$ and a density of 106.2 kg m$^{-3}$.
(ii) Particles of catalyst of diameter 0.1 mm and density 2510 kg m$^{-3}$ are fluidised in an oil of density 800 kg m$^{-3}$ and viscosity $2.85 \times 10^{-3}$ kg m$^{-1}$ s$^{-1}$. The vessel used is cylindrical, with an inside diameter of 0.63 m, and the height of the catalyst bed under minimum fluidising conditions is 4.01 m. The volumetric flow rate of the oil under minimum fluidising conditions is $2.04 \times 10^{-5}$ m$^3$ s$^{-1}$.

Acceleration due to gravity = 9.81 m s$^{-2}$

*Solution*

(i) The value of the Froude number under minimum fluidising conditions $(Fr_{mf})$ is often used as a criterion for predicting whether particulate or aggregative fluidisation occurs. However the product of the following four dimensionless groups is considered to be a more reliable guide (ref. 1 and 2) and will be used for the present problem.

$$(Fr_{mf})(Re_{p,mf})\left(\frac{\rho_s - \rho}{\rho}\right)\left(\frac{L_{mf}}{d_t}\right)$$

$Re_{p,mf}$ is the Reynolds number, under minimum fluidising conditions, based on the particle diameter and the density, viscosity and superficial linear velocity under minimum fluidising conditions of the fluid. $L_{mf}$ is the height of the fluidised bed at minimum fluidising conditions, and $d_t$ is the diameter of the fluidised bed.

If the above product of dimensionless groups is less than 100, the fluidisation is expected to be particulate or smooth. Should the product be greater than 100, aggregative or bubbling fluidisation will probably occur.

For the present problem,

$$Re_{p,mf} = d\rho U_{mf}/\mu$$

$$= \frac{0.3 \times 10^{-3} \times 106.2 \times 2.10 \times 10^{-2}}{2.21 \times 10^{-5}} = 30.3$$

$$Fr_{mf} = \frac{U_{mf}^2}{dg}$$

$$= \frac{(2.10 \times 10^{-2})^2}{0.3 \times 10^{-3} \times 9.81} = 0.15$$

$$\frac{\rho_s - \rho}{\rho} = \frac{2600 - 106.2}{106.2} = 23.5$$

$$\frac{L_{mf}}{d_t} = \frac{3.25}{0.95} = 3.42$$

Thus $(Fr_{mf})(Re_{p,mf})\left(\frac{\rho_s - \rho}{\rho}\right)\left(\frac{L_{mf}}{d_t}\right) = 0.15 \times 30.3 \times 23.5 \times 3.42$

$$= 365$$

This is greater than 100, so in this system aggregative or bubbling fluidisation will occur.

(ii) The volumetric flow rate of the fluid = $2.04 \times 10^{-5}$ m$^3$ s$^{-1}$ at minimum fluidising conditions.

$$U_{mf} = \frac{Q_{mf}}{A}$$

$$A = \frac{\pi d_t^2}{4}$$

$$= \frac{\pi \times 0.63^2}{4} = 0.312 \text{ m}^2$$

Thus $U_{mf} = \frac{2.04 \times 10^{-5}}{0.312} = 6.54 \times 10^{-5}$ m s$^{-1}$

Hence
$$Re_{p,mf} = \frac{d\rho U_{mf}}{\mu}$$
$$= \frac{0.1 \times 10^{-3} \times 800 \times 6.54 \times 10^{-5}}{2.85 \times 10^{-3}} = 1.84 \times 10^{-3}$$

and
$$Fr_{mf} = \frac{U_{mf}^2}{dg}$$
$$= \frac{(6.54 \times 10^{-5})^2}{0.1 \times 10^{-3} \times 9.81} = 4.36 \times 10^{-6}$$

$$\frac{\rho_s - \rho}{\rho} = (2510 - 800)/800 = 2.14$$

$$\frac{L_{mf}}{d_t} = \frac{4.01}{0.63} = 6.37$$

Therefore
$$(Fr_{mf})(Re_{p,mf})\left(\frac{\rho_s - \rho}{\rho}\right)\left(\frac{L_{mf}}{d_t}\right) = 4.36 \times 10^{-6} \times 1.84 \times 10^{-3} \times 2.14 \times 6.37$$
$$= 1.09 \times 10^{-7}$$

This is much less than 100, so that the fluidisation in this system occurs in the particulate or smooth mode.

## 11.1.2 The calculation of minimum flow rates, maximum flow rates and pressure drops in fluidised systems in which the flow is streamline

*Problem*

A catalyst which consists of spherical particles of diameter = $50 \times 10^{-6}$ m and density 1850 kg m$^{-3}$ is to be contacted with a liquid hydrocarbon of density 880 kg m$^{-3}$ and viscosity $2.75 \times 10^{-3}$ kg m$^{-1}$ s$^{-1}$ in a fluidised bed reactor. The depth of packing in the reactor when the fluid is not flowing is 1.37 m.

(i) At what superficial linear flow rate of hydrocarbon will fluidisation occur?
(ii) At what superficial linear flow rate of hydrocarbon will the catalyst particles begin to flow out of the fluidised bed with the hydrocarbon?
(iii) What is the pressure drop across the fluidised bed under minimum fluidisation conditions?

The void fraction of the catalyst bed at the point of minimum fluidisation was shown by experiment to be 0.45.

Acceleration due to gravity = 9.81 m s$^{-2}$.

## Solution

In a fluidised bed the force acting upwards on the particles must be equal to the effective weight of the particles in the fluid stream. It follows that if $L_{mf}$ is the bed height at minimum fluidisation, $A$ its cross-sectional area and $e_{mf}$ the void fraction at minimum fluidisation, then

$$(-\Delta P)A = g[\rho_s(1-e_{mf})L_{mf} A - \rho(1-e_{mf})L_{mf} A]$$

Thus
$$-\Delta P = g(1-e_{mf})(\rho_s - \rho)L_{mf} \qquad (11.1)$$

If flow conditions within the fluidised bed are streamline, at minimum fluidisation the superficial velocity through the bed may be estimated from the following equation (ref. 3)

$$U_{mf} = \frac{1}{K''} \frac{(e_{mf})^3}{S^2(1-e_{mf})^2} \frac{1}{\mu} \frac{(-\Delta P)}{L_{mf}} \qquad (11.2)$$

where $K''$ is usually assumed to be 5.0 and $S$ is the surface area per unit volume of the particles in the packing. For spheres

$$S = \frac{4\pi(d/2)^2}{4/3\pi(d/2)^3} = \frac{6}{d} \qquad (11.3)$$

The elimination of $-\Delta P$, $L_{mf}$ and $S$ from equations (11.1), (11.2) and (11.3) gives

$$U_{mf} = 5.5 \times 10^{-3} \frac{(e_{mf})^3}{1-e_{mf}} d^2 \frac{(\rho_s - \rho)}{\mu} g \qquad (11.4)$$

which may be used to determine the superficial velocity, $U_{mf}$, at minimum fluidisation (ref. 4), provided the flow is streamline.

Substitution of known data into equation (11.4) gives

$$U_{mf} = 5.5 \times 10^{-3} \frac{0.45^3 \times (50 \times 10^{-6})^2 (1850-880) \times 9.81}{(1-0.45) \times 2.75 \times 10^{-3}} \text{ m s}^{-1}$$

$$= 7.88 \times 10^{-6} \text{ m s}^{-1}$$

To check that the flow is in fact streamline, the criterion is that the Reynolds number, $Re$, defined as

$$Re_1 = \frac{\rho u}{S(1-e)\mu}$$

should be less than about 2 (ref. 5). In this problem, as $S = 6/d$, at incipient fluidisation we have that

$$Re_1 = \frac{U_{mf}\rho}{(6/d)(1-e_{mf})\mu}$$

$$= \frac{7.88 \times 10^{-6} \times 880}{[6/(50 \times 10^{-6})](1-0.45)(2.75 \times 10^{-3})}$$

$$= 3.8 \times 10^{-5}$$

The Reynolds number is thus less than 2, the flow is streamline, and so the use of equation (11.2) is justified.

Fluidisation therefore first occurs when the superficial linear flow rate of the hydrocarbon is $7.88 \times 10^{-6}$ m s$^{-1}$.

(ii) The particles in the fluidised bed will begin to flow out of the bed with the liquid when the liquid velocity is equal to the terminal settling velocity, $U_t$, for the particles.

If it is assumed that the Reynolds number for settling particles is in the streamline range for this situation, then Stokes' law can be used to determine the terminal settling velocity.

In this case,
$$U_t = \frac{d^2 g(\rho_s - \rho)}{18\mu}$$

$$= \frac{(50 \times 10^{-6})^2 (9.81)(1850 - 880)}{18 \times 2.75 \times 10^{-3}} \text{ m s}^{-1}$$

$$= 4.81 \times 10^{-4} \text{ m s}^{-1}$$

The criterion for the use of Stokes' law is that the Reynolds number, $Re_p$, must be less than 0.2, where $Re_p$ is given by (ref. 6, 7 and 8)

$$Re_p = \frac{d\rho U_t}{\mu}$$

In the present case

$$Re_p = \frac{50 \times 10^{-6} \times 880 \times 4.81 \times 10^{-4}}{2.75 \times 10^{-3}}$$

$$= 7.7 \times 10^{-3}$$

The use of Stokes' law is therefore justified and catalyst particles will begin to flow out of the fluidised bed when the hydrocarbon has a superficial linear velocity of $4.8 \times 10^{-4}$ m s$^{-1}$.

(iii) The pressure drop across the fluidised bed at the point of minimum fluidisation can be estimated using equation (11.1), provided it is assumed that the depth of the packing in the bed when the fluid does not flow is equal to the bed height at minimum fluidisation.

If one makes this assumption, then substitution of known data into equation (11.1) gives:

$$-\Delta P = 9.81(1 - 0.45)(1850 - 880)(1.37) \text{ N m}^{-2}$$

$$= 7170 \text{ N m}^{-2}$$

The pressure drop across the fluidised bed under minimum fluidising conditions is thus $7.17$ kN m$^{-2}$.

## 11.1.3 The calculation of flow rates in fluidised beds, when flow is not streamline, using the Ergun equation. Flow rates for the fluidisation of particles with a distribution of sizes

*Problem*

A sample of almost spherical particles of mineral ore has a particle size distribution as given in Table 11.1.

**Table 11.1**

| Weight % of material of diameter less than $d$ | $d$ (mm) |
|---|---|
| 0 | 0.40 |
| 8 | 0.50 |
| 20 | 0.56 |
| 40 | 0.62 |
| 60 | 0.68 |
| 80 | 0.76 |
| 90 | 0.84 |
| 100 | 0.94 |

The sample is to be fluidised in air at 15 °C and $1.013 \times 10^5$ N m$^{-2}$ in a cylindrical vessel 30 cm in diameter and 1.65 m in height.
Determine:
(i) the air flow rate at which fluidisation first occurs and
(ii) the air flow rate at which the smallest particles will be carried out of the fluidised bed.

Density of the ore sample = 1980 kg m$^{-3}$
Density of the air = 1.218 kg m$^{-3}$
Viscosity of air = $1.73 \times 10^{-5}$ kg m$^{-1}$ s$^{-1}$
Acceleration due to gravity = 9.81 m s$^{-2}$
Void fraction at incipient fluidisation for the ore sample = 0.40

*Solution*

(i) The problem involves the fluidisation of moderately heavy particles in a gas stream. It is therefore likely that the Reynolds number, $Re_1$, as defined by

$$Re_1 = \frac{U_{mf} \rho}{S(1 - e_{mf}) \mu}$$

will be outside of the range to which the Carman–Kozeny equation applies and the Ergun equation, viz.

$$\frac{-\Delta P}{L_{mf}} = 150\frac{(1-e_{mf})^2}{e_{mf}^3}\frac{\mu U_{mf}}{d^2} + 1.75\frac{(1-e_{mf})}{e_{mf}^3}\frac{\rho U_{mf}^2}{d} \qquad (11.5)$$

should be used (ref. 9, 10, 11 and 12).

As was indicated for Example 11.1.2, a force balance on particles in a fluidised bed gives the equation

$$-\Delta P = g(1-e_{mf})(\rho_s - \rho)L_{mf} \qquad (11.6)$$

The elimination of $(-\Delta P)$ and $L_{mf}$ from equations (11.5) and (11.6) results in the following equation from which the superficial air flow rate at incipient fluidisation, $U_{mf}$, can be calculated (ref. 9, 10 and 11).

$$(1-e_{mf})(\rho_s-\rho)g = \frac{150(1-e_{mf})^2 \mu U_{mf}}{e_{mf}^3 d^2} + \frac{1.75(1-e_{mf})\rho U_{mf}^2}{e_{mf}^3 d}$$

or

$$(\rho_s-\rho)g = \frac{150(1-e_{mf})\mu U_{mf}}{e_{mf}^3 d^2} + \frac{1.75\rho U_{mf}^2}{e_{mf}^3 d} \qquad (11.7)$$

However equation (11.7) applies as written to spherical particles of uniform size. For particles which have a distribution of sizes, the particle diameter, $d$, should be replaced by the surface mean particle diameter, $\bar{d}$, where $\bar{d}$ can be determined from

$$\bar{d} = \frac{1}{\Sigma(x/d)_i}$$

in which $x$ is the mass fraction of particles of diameter $= d$  (ref. 13).

For the present problem, the surface mean particle diameter is calculated from the particle size distribution data as shown in Table 11.2.

**Table 11.2**

| Diameter range (mm) | Average diameter (mm) | Mass fraction in the interval | $\frac{x_i}{d_i}$ (mm$^{-1}$) |
|---|---|---|---|
| 0.40–0.50 | 0.45 | 0.08 | 0.1778 |
| 0.50–0.56 | 0.53 | 0.12 | 0.2264 |
| 0.56–0.62 | 0.59 | 0.20 | 0.3390 |
| 0.62–0.68 | 0.65 | 0.20 | 0.3077 |
| 0.68–0.76 | 0.72 | 0.20 | 0.2778 |
| 0.76–0.84 | 0.80 | 0.10 | 0.1250 |
| 0.84–0.94 | 0.89 | 0.10 | 0.1124 |

$$\Sigma\left(\frac{x_i}{d_i}\right) = 1.566 \text{ mm}^{-1}$$

whence
$$\bar{d} = \frac{1}{\Sigma\left(\frac{x_i}{d_i}\right)} = 0.639 \text{ mm}$$

Substitution of $\bar{d}$ for $d$ in equation (11.7), and the other known data, gives

$$(1980 - 1.218)9.81 = \frac{150(1-0.4)(1.73 \times 10^{-5})U_{mf}}{0.4^3(0.639 \times 10^{-3})^2} + \frac{1.75 \times 1.218 U_{mf}^2}{0.4^3 \times 0.639 \times 10^{-3}}$$

or
$$19\,412 = 59\,580 U_{mf} + 52\,120 U_{mf}^2$$

If this equation is solved by the formula for the solution of a quadratic equation, the positive root gives $U_{mf} = 0.26 \text{ m s}^{-1}$.

The superficial linear air flow rate at which fluidisation first occurs is therefore $0.26 \text{ m s}^{-1}$.

(ii) The ore particles of smallest diameter will be the first to flow out of the bed with increasing air velocity, and this will again occur when the air velocity equals the terminal settling velocity, $U_t$, of these particles. The problem is therefore to determine the terminal settling velocity of ore particles with a diameter of 0.4 mm.

For a spherical particle, the settling velocity can be determined using the table of $\log Re$ as a function of $\log(R'/\rho U_t^2)Re^2$ (ref. 14) or from charts (ref. 15 and 16).

Thus
$$(R'/\rho U_t^2)Re^2 = (2d^3/3\mu^2)\rho(\rho_s - \rho)g$$

$$= \frac{2 \times (0.4 \times 10^{-3})^3 (1.218)(1980 - 1.218)9.81}{3(1.75 \times 10^{-5})^2}$$

$$= 3294$$

$$\log_{10} 3294 = 3.518$$

From Table 3.2 of ref. 14 we obtain

$$\log_{10} Re = 1.853$$
$$Re = 71.29$$

Since
$$Re = \frac{U_t \rho d}{\mu}$$

$$U_t = Re\,\mu/\rho d$$
$$= \frac{71.29 \times 1.73 \times 10^{-5}}{1.218 \times 0.4 \times 10^{-3}} \text{ m s}^{-1}$$
$$= 2.53 \text{ m s}^{-1}$$

224  Problems in fluid flow

The smallest particles will therefore be carried out of the fluidised bed when the superficial linear air flow rate reaches $2.53 \text{ m s}^{-1}$.

## 11.1.4 The estimation of vessel diameters and heights for fluidisation operations

*Problem*

A bed containing 102 kg of spherical catalyst particles of density $1500 \text{ kg m}^{-3}$ and diameter 0.2 mm is to be fluidised in a vertical, cylindrical reaction vessel using a liquid hydrocarbon. The hydrocarbon has to be processed continuously at a rate of 125 kg per hour. It has a density of $852 \text{ kg m}^{-3}$ and a viscosity of $1.92 \times 10^{-3} \text{ kg m}^{-1} \text{s}^{-1}$.

The volume fraction of liquid in the bed was estimated by filling a $250 \text{ cm}^3$ measuring cylinder to the $200 \text{ cm}^3$ mark with catalyst, and then adding liquid until the $250 \text{ cm}^3$ mark was reached, taking care not to trap air bubbles. It was found that $136.5 \text{ cm}^3$ of liquid was required for this.

If the flow rate of liquid in the fluidised bed is twice that needed to achieve incipient fluidisation, estimate
(i) the diameter required for the reaction vessel
(ii) the minimum depth of the fluidised bed, and
(iii) the depth of the fluidised bed under operating conditions.

*Solution*

(i) Use will be made of the equations:

$$U_{mf} = \frac{1}{K'' S^2} \cdot \frac{e_{mf}^3}{(1-e_{mf})^2} \cdot \frac{1}{\mu} \cdot \frac{(-\Delta P)}{L_{mf}} \tag{11.8}$$

and

$$-\Delta P = (1-e_{mf})(\rho_s - \rho)L_{mf}g \tag{11.9}$$

(ref. 1, 2 and 3) as in example 11.1.2 to determine $U_{mf}$, the superficial velocity through the vessel under conditions of minimum fluidisation. The superficial velocity at operating conditions will then be calculated, and thence the mass flow rate of liquid per unit cross section of the vessel. From this and the given liquid processing rate, the area, and therefore the diameter, of the vessel can be estimated.

Elimination of $-\Delta P$ from equations (11.8) and (11.9), and substitution of $K'' = 5.0$ and $S = 6/d$, as in example 11.1.2, gives

$$U_{mf} = 5.5 \times 10^{-3} \frac{e_{mf}^3}{(1-e_{mf})} \cdot \frac{d^2 (\rho_s - \rho)g}{\mu} \tag{11.10}$$

To calculate the void fraction, $e_{mf}$, it is noted that when the total volume of catalyst plus liquid in the measuring cylinder was $200 \text{ cm}^3$, the volume of liquid added = 136.5 − 50 = 86.5 cm³. It follows that

$$e_{mf} = \frac{86.5}{200} = 0.43$$

Substituting data into equation (11.10) gives

$$U_{mf} = 5.5 \times 10^{-3} \times \frac{0.43^3}{1-0.43} \times \frac{(0.2 \times 10^{-3})^2(1500-852)(9.81)}{1.92 \times 10^{-3}} \text{ m s}^{-1}$$

$$= 1.016 \times 10^{-4} \text{ m s}^{-1}$$

The superficial velocity under operating conditions, $U$, is given by

$$U = 2U_{mf}$$

$$= 2 \times 1.016 \times 10^{-4} = 2.032 \times 10^{-4} \text{ m s}^{-1}$$

The mass flow rate of fluid per unit cross sectional area of the vessel is given by

$$G = U\rho$$

$$= 2.032 \times 10^{-4} \times 852 = 0.173 \text{ kg m}^{-2} \text{ s}^{-1}$$

The actual mass flow rate of fluid equals 125 kg h⁻¹.

Thus, mass flow rate $= \dfrac{125}{3600} \text{ kg s}^{-1}$

$$= GA$$

where $A$ is the cross sectional area.

Therefore $\quad A = \dfrac{125}{3600G} \text{ m}^2$

$$= \frac{125}{3600 \times 0.173} = 0.201 \text{ m}^2$$

As $\quad A = \dfrac{\pi d_t^2}{4}$

$$d_t = \sqrt{\frac{4A}{\pi}} = \sqrt{\frac{4 \times 0.201}{\pi}} \text{ m}$$

$$= 0.51 \text{ m}$$

To justify the use of equation (11.8) the Reynolds number, $Re_1$, defined as

$$Re_1 = \frac{U\rho}{S(1-e)\mu}$$

should have a value less than about 2 (ref. 3). In the present case, as $S = 6/d$, at incipient fluidisation

$$Re_1 = \frac{1.016 \times 10^{-4} \times 852}{[6/(0.2 \times 10^{-3})] \times (1-0.43) \times 1.92 \times 10^{-3}}$$

$$= 2.64 \times 10^{-3}$$

so the use of equation (11.8) is justified.

Thus the vessel should have a diameter of 0.51 m.

(ii) If $L$ is the depth of the fluidised bed when its void fraction is $e$, and $L_{mf}$ the depth at incipient fluidisation, then (ref. 16),

$$L = L_{mf} \frac{1-e_{mf}}{1-e} \qquad (11.11)$$

Now the total volume $V$ of solid catalyst particles in the vessel is given by

$$V = \frac{\text{mass of solids}}{\text{density of solids}}$$

$$= \frac{102 \text{ kg}}{1500 \text{ kg m}^{-3}} = 0.068 \text{ m}^3$$

The depth $L_0$ that this volume of solids would occupy in the vessel if the void volume were zero is given by

$$L_0 = \frac{V}{A}$$

$$= \frac{0.068}{0.201} = 0.338 \text{ m}$$

Substituting $e = 0$ and $L = L_0 = 0.338$ m into equation (11.11), followed by rearrangement, gives the depth, $L_{mf}$, of the bed at incipient fluidisation.

Thus
$$L_{mf} = \frac{0.338}{1-0.43} = 0.59 \text{ m}$$

[Note that $A = 0.201$ m$^2$ and $e_{mf} = 0.43$ have been determined in section (i) of this problem.]

The minimum depth of the fluidised bed is therefore 0.59 m.

(iii) Equation (11.11) can also be used to determine the depth of the fluidised bed under operating conditions provided the void fraction $e$ of the bed under these operating conditions is first determined. The void fraction can be determined using equation (11.10), by substituting the operating linear velocity for $U_{mf}$, and then solving for $e$. Thus we have

$$2.032 \times 10^{-4} = 5.5 \times 10^{-3} \frac{e^3}{1-e} \times \frac{(0.2 \times 10^{-3})^2 (1500-852)(9.81)}{1.92 \times 10^{-3}}$$

whence
$$e = 0.514$$

($e$ is found by trial and error)

Substituting $e = 0.514$, and $L_{mf} = 0.59$ m, $e_{mf} = 0.43$ (previously determined) into equation (11.11) gives,

$$L = 0.59 \left( \frac{1 - 0.43}{1 - 0.514} \right) = 0.69 \text{ m}$$

The depth of the fluidised bed under operating conditions is therefore 0.69 m.

## 11.1.5  Power required for pumping in fluidised beds

*Problem*

Glass spheres which have a diameter of 1.0 mm are to be fluidised by air using apparatus which is shown diagrammatically in Figure 11.1.

The air enters the blower at a pressure of $1.013 \times 10^5$ N m$^{-2}$, and leaves from the top of the fluidised bed at the same pressure. Prior to fluidisation the

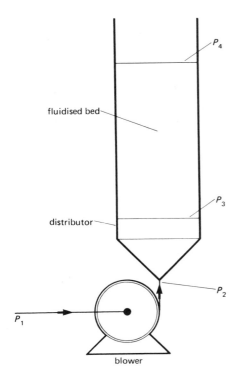

**Figure 11.1**

228  Problems in fluid flow

bed is 7.32 m high, has a diameter of 1.02 m, and a voidage of 0.48. If the superficial air velocity entering the bed is 0.015 m s$^{-1}$, the pressure drop through the distributor is 10% of the pressure drop through the bed, and the blower is 70% efficient, determine the power which has to be supplied to the blower to fluidise the bed.

Density of the glass = 2280 kg m$^{-3}$
The density of the air entering the bottom of the fluidised bed = 1.204 kg m$^{-3}$
$\gamma$, the ratio of heat capacities, $C_p/C_v$, for the air = 1.4
Acceleration due to gravity = 9.81 m s$^{-2}$

*Solution*

From Figure 11.1 it is seen that the power required to supply air to the fluidised bed at pressure $P_3$ is the power needed to compress air from $P_1$ to $P_2$. This may be determined by doing a mechanical energy balance between positions 1 and 2, that is, across the blower.

Thus

$$\Delta\left(\frac{U^2}{2\alpha}\right) + g\Delta z + \int_1^2 v\,dP + W_s + F = 0 \qquad (11.12)$$

as given in ref. 17 to 20.

Now since the blower is not cooled, the fluid is compressed adiabatically. Further, if it is assumed that the kinetic energy, potential energy and friction loss terms of equation (11.12) can be neglected in this situation, then the shaft work $(-W_s)$ done on the system by the blower is given by

$$-W_s = \int_{P_1}^{P_2} v\,dP$$

With the additional assumption that the air in this case behaves as an ideal gas, it follows that (ref. 21 and 22)

$$-W_s = \frac{\gamma}{\gamma - 1}(P_2 v_2 - P_1 v_1) \qquad (11.13)$$

The blower has an efficiency, $\eta = 0.70$. The energy $(-W_p)$ supplied to the pump is therefore given by

$$-W_p = \frac{-W_s}{\eta} = \frac{-W_s}{0.70}$$

Substitution for $W_s$ in equation (11.13) gives

$$-W_p = \frac{1}{\eta} \cdot \frac{\gamma}{\gamma - 1} \cdot (P_2 v_2 - P_1 v_1)$$

By inserting data known at this stage in the above equation we have

$$-W_p = \frac{1}{0.7} \times \frac{1.4}{(1.4-1)} \left( P_2 v_2 - \frac{1.013 \times 10^5}{1.204} \right) \text{J kg}^{-1} \quad (11.14)$$

To evaluate $P_2$, a force balance across the fluidised bed gives

$$-\Delta P = g(1-e)(\rho_s - \rho)L \quad \text{(see example 11.1.2)} \quad (11.15)$$

Thus
$$-\Delta P = 9.81\,(1-0.48)(2280-1.204)(7.32)\,\text{N m}^{-2}$$
$$= 85\,090\,\text{N m}^{-2}$$
$$= P_3 - P_4$$

Therefore
$$P_3 = P_4 + (-\Delta P)$$
$$= (1.013 \times 10^5) + 85\,090\,\text{N m}^{-2}$$
$$= 186\,400\,\text{N m}^{-2}$$

From the statement that the pressure drop through the distributor is equal to 10% of the pressure drop through the bed, (ref. 23)

$$P_2 - P_3 = 0.1\,(85\,090) = 8509\,\text{N m}^{-2}$$

Therefore as
$$P_2 = 8509 + P_3,$$
$$P_2 = 8509 + 186\,400\,\text{N m}^{-2}$$
$$= 194\,900\,\text{N m}^{-2}$$

The value of $v_2$ can now be determined. As it has been assumed that the compression of the air by the blower is adiabatic, then

$$P_1 v_1^\gamma = P_2 v_2^\gamma$$

$$v_2 = \left( \frac{P_1 v_1^\gamma}{P_2} \right)^{\frac{1}{\gamma}}$$

$$= \left( \frac{1.013}{1.949} \times 0.831^{1.4} \right)^{\frac{1}{1.4}} \text{m}^3\,\text{kg}^{-1}$$

since
$$v_1 = \frac{1}{\rho_1} = \frac{1}{1.204} = 0.831\,\text{m}^3\,\text{kg}^{-1}$$

So
$$v_2 = 0.520\,\text{m}^3\,\text{kg}^{-1}$$

The energy per unit mass of air ($-W_p$) required for the blower is now given by substituting the values calculated above for $P_2$ and $v_2$ into equation (11.14). Thus

$$-W_p = \frac{1}{0.7} \times \frac{1.4}{(1.4-1)} \left[ (1.949 \times 10^5 \times 0.520) - \frac{(1.013 \times 10^5)}{1.204} \right] \text{J kg}^{-1}$$
$$= 86\,100\,\text{J kg}^{-1}$$

Now the mass flow rate of the air through the blower equals the mass flow rate of air entering the fluidised bed. We have no information from which to calculate the temperature of the air entering the bed, but assuming isothermal flow of air from the blower through the distributor, which seems a reasonable approximation (ref. 24), we have that

$$P_2 v_2 = P_3 v_3$$

where $v_3$ is the specific volume of the air at the bottom of the fluidised bed. Therefore,

$$\begin{aligned}\rho_3 &= 1/v_3 \\ &= P_3/P_2 v_2 \\ &= (1.864 \times 10^5)/(1.949 \times 10^5 \times 0.520) \, \text{kg m}^{-3} \\ &= 1.84 \, \text{kg m}^{-3}\end{aligned}$$

Now the mass flow rate of air entering the fluidised bed

$$\begin{aligned}&= UA\rho_3 \\ &= 0.015 \times \pi \times 0.51^2 \times 1.84 = 0.0226 \, \text{kg s}^{-1}\end{aligned}$$

The power to be supplied to the blower therefore

$$= 0.0226 \times 86\,100 \, \text{J s}^{-1} = 1946 \, \text{watts}$$

It should be noted that there are many assumptions in the above treatment, and this answer is therefore only approximate.

## 11.1.6 The wall effect in fluidised beds using narrow columns

*Problem*

In a laboratory experiment spherical catalyst particles of diameter 1.8 mm were fluidised in water using a glass column which had an inside diameter of 12.7 mm. The volumetric flow rate of the water at the point of incipient fluidisation was found to be 2.306 cm³ s⁻¹.

Estimate the superficial linear flow rate of water at incipient fluidisation of the same catalyst particles packed to the same depth, but in a glass column which has an inside diameter of 150 cm, if all other conditions remain unchanged.

*Solution*

In a packed bed the particles do not pack as closely near the wall of the column as they usually do at the centre. The resistance to flow due to the packing may therefore be greater at the centre of the bed than it is near the wall. This effect, termed the 'wall effect', will be most noticeable in very narrow columns.

Coulson (ref. 25 and 26) has used an experimentally determined correction factor, $f_w$, to compensate for the wall effect.

Thus
$$f_w = \left(1 + 0.5 \frac{S_c}{S}\right)^2 \qquad (11.16)$$

where $S_c$ is the surface area of the container per unit volume of bed, and $S$ is the surface area of the packing per unit volume of packing.

Thus if the column diameter $= d_t$, and the depth of the packed bed $= L$, the surface area of the container $= \pi d_t L$ and the volume of the packed bed $= \pi (d_t/2)^2 L$

Hence
$$S_c = \frac{\pi d_t L}{\pi \left(\dfrac{d_t}{2}\right)^2 L} = \frac{4}{d_t}$$

For the present problem, for the smaller column,
$$S_c = \frac{4}{12.7 \times 10^{-3}} = 315 \text{ m}^{-1}$$

Furthermore, for spherical particles,
$$S = \frac{4\pi \left(\dfrac{d}{2}\right)^2}{\dfrac{4}{3}\pi \left(\dfrac{d}{2}\right)^3} = \frac{6}{d}$$

$$= \frac{6}{1.8 \times 10^{-3}} = 3333 \text{ m}^{-1}$$

For the smaller glass column therefore,
$$f_w = \left(1 + 0.5 \frac{315}{3333}\right)^2$$
$$= 1.1$$

The volumetric flow rate of water at incipient fluidisation was 2.306 cm$^3$ s$^{-1}$ in the small column, or $2.306 \times 10^{-6}$ m$^3$ s$^{-1}$. The superficial linear velocity at incipient fluidisation, $U_{mf}$, is thus given by

$$U_{mf} = \frac{Q_{mf}}{\pi \left(\dfrac{d_t}{2}\right)^2}$$

$$= \frac{2.306 \times 10^{-6}}{\pi \left(\dfrac{12.7 \times 10^{-3}}{2}\right)^2} \text{ m s}^{-1}$$

$$= 0.0182 \text{ m s}^{-1}$$

The minimum fluidising velocity found using the smaller glass column, when corrected for the wall effect using Coulson's equation (11.16), is thus

$$= 1.1 \times 0.0182 = 0.020 \text{ m s}^{-1}$$

Now for the column which has a diameter of 150 cm or 1.50 m,

$$S_c = \frac{4}{d_t}$$

$$= \frac{4}{1.50} = 2.67 \text{ m}^{-1}$$

As before, $\quad S = 3333 \text{ m}^{-1}$

(as the same catalyst particles are to be used.)

Thus $\quad f_w = \left(1 + 0.5 \, \frac{2.67}{3333}\right)^2$

$$= 1.00$$

No correction of the minimum fluidising velocity for the wall effect is therefore needed when the larger glass column is used to fluidise the catalyst particles.

The superficial linear flow rate of water at incipient fluidisation using this column is therefore $0.018 \text{ m s}^{-1}$.

## 11.1.7 The effect of particle size on the ratio of the terminal velocity of particles in a fluid to the minimum fluidisation velocity

### Problem

An ore contains a very wide particle size distribution, including both very large and very small particles. The ore is fractionated into narrow size ranges, and each fraction is to be fluidised separately in air. All particles are very nearly spherical. Predict the maximum value and the minimum value of the ratio $U_t/U_{mf}$, where $U_t$ is the terminal velocity of a particle of a given size in air, and $U_{mf}$ is the minimum fluidising velocity expected when the fraction containing only particles of the same given size is fluidised in air. The void fraction at incipient fluidisation for all particles is 0.4.

### Solution

A convenient starting point for this problem is the Ergun equation (ref. 8, 9, 10 and 11) which when applied to a fluidised bed at the point of incipient fluidisation may be written:

$$-\frac{\Delta P}{L_{mf}} = 150 \frac{(1-e_{mf})^2}{e_{mf}^3} \frac{\mu U_{mf}}{d^2} + 1.75 \frac{(1-e_{mf})}{e_{mf}^3} \rho \frac{U_{mf}^2}{d} \quad (11.17)$$

A force balance on the particles in a fluidised bed, at the point of incipient fluidisation, gives (see example 11.1.3)

$$-\Delta P = g(1-e_{mf})(\rho_s - \rho)L_{mf} \quad (11.18)$$

The elimination of $-\Delta P/L_{mf}$ from equations (11.17) and (11.18) again gives the following equation,

$$(\rho_s - \rho)g = \frac{150(1-e_{mf})\mu U_{mf}}{e_{mf}^3 d^2} + \frac{1.75\rho U_{mf}^2}{e_{mf}^3 d} \quad (11.19)$$

Now for very large particles, the first term on the right-hand side of this equation becomes small compared with the second, and so can be neglected (ref. 27).

Thus for large particles

$$(\rho_s - \rho)g = \frac{1.75\rho U_{mf}^2}{e_{mf}^3 d}$$

or

$$U_{mf}^2 = \frac{d(\rho_s - \rho)g e_{mf}^3}{1.75\rho} \quad (11.20)$$

This equation holds for $Re_p > 1000$ (ref. 27).

The terminal velocity, $U_t$, for a particle settling in a fluid is given by

$$U_t^2 = \frac{4gd(\rho_s - \rho)}{3\rho C_D} \quad (11.21)$$

where the drag coefficient, $C_D$, for $500 < Re_p < 200\,000$, equals 0.43. Substituting for $C_D$ in equation (11.21) then gives

$$U_t^2 = \frac{4gd(\rho_s - \rho)}{3\rho(0.43)}$$

i.e.

$$U_t^2 = \frac{3.1\,gd(\rho_s - \rho)}{\rho} \quad (11.22)$$

From equations (11.22) and (11.20) we find that

$$\frac{U_t^2}{U_{mf}^2} = \frac{3.1\,gd(\rho_s - \rho)/\rho}{d(\rho_s - \rho)g e_{mf}^3/1.75\rho}$$

$$= \frac{3.1 \times 1.75}{e_{mf}^3} = \frac{3.1 \times 1.75}{0.43^3}$$

$$= 84.8$$

Thus
$$\frac{U_t}{U_{mf}} = 9.21$$

Therefore for the largest particles, for which $Re_p > 500$, the ratio of the terminal velocity of a particle to the minimum fluidising velocity equals 9.21.

For the very small particles it is the second term on the right-hand side of equation (11.19) that can be neglected, and so for the smallest ore particles we have

$$(\rho_s - \rho)g = \frac{150(1 - e_{mf})\mu U_{mf}}{e_{mf}^3 d^2}$$

This equation is valid for $Re_p < 20$ (ref. 27). Rearrangement gives, for the smallest particles,

$$U_{mf} = \frac{(\rho_s - \rho)g\, e_{mf}^3 d^2}{150(1 - e_{mf})\mu} \quad (11.23)$$

Furthermore, for the settling of particles where $Re_p < 0.2$, i.e. the Stokes law range,

$$C_D = \frac{24}{Re_p}$$

$$= \frac{24\mu}{d U_t \rho}$$

Substitution of this equation for $C_D$ into equation (11.21) gives

$$U_t^2 = \frac{4gd^2(\rho_s - \rho)U_t}{72\mu}$$

or
$$U_t = \frac{g(\rho_s - \rho)d^2}{18\mu} \quad (11.24)$$

From equations (11.23) and (11.24) it follows that for the smallest particles,

$$\frac{U_t}{U_{mf}} = \frac{g(\rho_s - \rho)d^2/18\mu}{(\rho_s - \rho)g\, e_{mf}^3 d^2/[150(1 - e_{mf})\mu]}$$

$$= \frac{150(1 - e_{mf})}{18\, e_{mf}^3}$$

For the ore particles, $e_{mf} = 0.4$

Therefore
$$\frac{U_t}{U_{mf}} = \frac{150(1 - 0.4)}{18 \times 0.4^3}$$

$$= 78.1$$

Thus we predict that the maximum value of the ratio $U_t/U_{mf}$ equals 78.1, and this holds for the smallest ore particles, provided that $Re_p < 0.2$. The minimum value of the ratio $U_t/U_m$ equals 9.21, and this is true for the largest ore particles, provided that $500 < Re_p < 200\,000$, since this is the range for which $C_D$ equals 0.43.

## 11.2 Student Exercises

1   A sample of ore which has a uniform particle diameter of $50 \times 10^{-6}$ m and a density of $4000$ kg m$^{-3}$ is fluidised in an organic liquid of density $790$ kg m$^{-3}$ and viscosity $1.30 \times 10^{-3}$ kg m$^{-1}$ s$^{-1}$. Assuming that the particles are spherical, derive an equation showing the relationship between the superficial liquid velocity through the fluidised bed and the volume fraction of liquid in the bed. The volume fraction of liquid in a closely packed bed was 0.42. Acceleration due to gravity $= 9.81$ m s$^{-2}$.

2   A fluidised bed reactor of diameter 0.875 m is to be used to fluidise anthracite particles of density $1760$ kg m$^{-3}$ with a liquid petroleum fraction of density $880$ kg m$^{-3}$. The maximum pressure drop that can be developed across the fluidised bed by the pump is $9162$ N m$^{-2}$.
(i) What is the maximum amount of anthracite that should be put into the reaction vessel?
(ii) If the depth of the fluidised bed, when operating under the above conditions, is 3.9 m, what would be the volume fraction of liquid in the column?

3   Table 11.3 gives a particle size analysis for a coal sample.

**Table 11.3**

| Mass % of sample of diameter less than $d$ | $d$ (mm) |
|---|---|
| 0 | 0.40 |
| 7 | 0.56 |
| 20 | 0.66 |
| 30 | 0.68 |
| 47 | 0.72 |
| 58 | 0.74 |
| 74 | 0.78 |
| 85 | 0.80 |
| 100 | 0.90 |

(i) Calculate the surface mean diameter from the above particle size distribution.

(ii) Calculate the air flow rate at which incipient fluidisation occurs with this sample.
Assume the coal particles are spherical.

Density of coal sample = 1750 kg m$^{-3}$
Density of air = 1.218 kg m$^{-3}$
Viscosity of air = $1.73 \times 10^{-5}$ kg m$^{-1}$ s$^{-1}$
Void fraction of coal in a loosely packed bed = 0.43
Acceleration due to gravity = 9.81 m s$^{-2}$

4  Butane gas is to be processed continuously in a fluidised bed reactor containing 22.0 kg of a catalyst which consists of spherical particles 4.4 mm in diameter. Under the specified operating conditions the gas flow rate is to be maintained at 560 m$^3$ per hour. Assuming that the superficial linear velocity through the fluidised bed is three times that required for minimum fluidisation, estimate the diameter of the reactor that should be used.

Density of butane at operating temperature = 2.50 kg m$^{-3}$
Viscosity of butane at operating temperature = $1.12 \times 10^{-5}$ kg m$^{-1}$ s$^{-1}$
Density of catalyst = 1410 kg m$^{-3}$
Void fraction of catalyst in static bed = 0.41
Acceleration due to gravity = 9.81 m s$^{-2}$

5  A loosely packed bed of particles of density 2600 kg m$^{-3}$ and diameter 1.0 mm has a bed depth of 3.02 m. When ethanol flows up through the bed the pressure drop at incipient fluidisation is 28 970 N m$^{-2}$.

If the liquid is changed to tetrachloroethylene, but the bed of particles is otherwise unchanged, what pressure drop would then be required to fluidise the bed?

Density of ethanol = 789 kg m$^{-3}$
Density of tetrachloroethylene = 1624 kg m$^{-3}$
Acceleration due to gravity = 9.81 m s$^{-2}$

6  The Galileo number ($Ga$) for a particle of density $\rho_s$ and diameter $d$ in a fluid of density $\rho$ and viscosity $\mu$ is defined as (ref. 13)

$$Ga = \frac{\rho(\rho_s - \rho)gd^3}{\mu^2}$$

(i) If $U_t$ is the terminal settling velocity of spherical catalyst particles in water, and $U_{mf}$ is the superficial linear velocity of the water flowing through a bed of catalyst particles at the point of incipient fluidisation, prepare a graph showing the relationship between the Galileo number and the ratio $U_t/U_{mf}$.

(ii) Comment briefly on the shape of the graph.

Density of water = 998 kg m$^{-3}$
Viscosity of water = 1.25 × 10$^{-3}$ kg m$^{-1}$ s$^{-1}$
Density of catalyst particles = 1950 kg m$^{-3}$
Void fraction of bed of catalyst particles at incipient fluidisation = 0.42
Acceleration due to gravity = 9.81 m s$^{-2}$

7  An ore sample is fluidised in a liquid hydrocarbon. The ore particles are spherical, and have a diameter of 0.10 mm and a density of 2550 kg m$^{-3}$. When a loosely packed bed of the particles was settled in the hydrocarbon the void fraction was found to be 0.43.

Estimate the superficial linear velocity required to make the bed expand to a volume equal to three times the volume of the loosely packed bed.

Density of hydrocarbon = 920 kg m$^{-3}$
Viscosity of hydrocarbon = 0.002 kg m$^{-1}$ s$^{-1}$
Acceleration due to gravity = 9.81 m s$^{-2}$

8  Sand is to be fluidised in a tank which has a diameter of 0.6 metres using air which is to flow at a superficial linear velocity of 0.40 m s$^{-1}$. The height of the loosely packed bed of sand prior to fluidisation is 2.75 metres, and the void fraction is 0.45. The air enters the bottom of the fluidised bed through a perforated plate distributor, and leaves the top of the bed at atmospheric pressure.

If the pressure drop across the distributor is 10 % of the pressure drop across the fluidised bed, and each hole in the distributor has a diameter of 1.0 mm, how many holes are required?

Atmospheric pressure = 1.013 × 10$^5$ N m$^{-2}$
Density of air = 1.22 kg m$^{-3}$
Viscosity of air = 1.70 × 10$^{-5}$ kg m$^{-1}$ s$^{-1}$
Density of the sand = 2650 kg m$^{-3}$
Acceleration due to gravity = 9.81 m s$^{-2}$

## 11.3  References

Four references are abbreviated as follows:

C and R (2) = J. M. Coulson and J. F. Richardson, Chemical Engineering, Vol. 2, 3rd edition, Pergamon, 1980

F et al. = A. S. Foust, L. A. Wenzel, C. W. Clump, L. Maus and L. B. Andersen, Principles of Unit Operations, 2nd edition, John Wiley and Sons, 1980

K and L = D. Kunii and O. Levenspiel, Fluidisation Engineering, John Wiley and Sons, 1969

238  Problems in fluid flow

McC, S & H = W. L. McCabe, J. C. Smith and P. Harriott, Unit Operations of Chemical Engineering, 4th edition, McGraw Hill, 1985

1. K and L, page 81
2. F et al., page 643
3. C and R (2), pages 129 and 130
4. C and R (2), page 233
5. C and R (2), page 89
6. McC, S & H, page 129
7. F et al., page 614
8. C and R (2), page 234
9. McC, S & H, page 149
10. F et al., page 642
11. K and L, page 66
12. K and L, page 67
13. C and R (2), page 97
14. C and R (2), page 98
15. J. F. Douglas, J. M. Gasiorek and J. A. Swaffield, Fluid Mechanics, Pitman, 1980, page 326
16. McC, S & H, page 152
17. J. M. Coulson and J. F. Richardson, Chemical Engineering, Vol. 1, 3rd edition, Pergamon, 1977, page 27
18. McC, S & H, page 68
19. J. F. Douglas, J. M. Gasiorek and J. A. Swaffield, Fluid Mechanics, Pitman, 1980, page 160
20. K and L, page 100
21. McC, S & H, page 188
22. J. M. Coulson and J. F. Richardson, Chemical Engineering, Vol. 1, 3rd edition, Pergamon, 1977, page 30
23. K and L, page 87
24. K and L, page 103
25. J. M. Coulson, *Trans. Inst. Chem. Eng.*, 1949 **27**, 237–57
26. C and R (2), page 132
27. K and L, page 73
28. K and L, page 90

## 11.4 Notation

| Symbol | Description | Unit |
|---|---|---|
| $A$ | cross-sectional area of bed or column | $m^2$ |
| $C_p$ | heat capacity at constant pressure per unit mass | $J\,kg^{-1}K^{-1}$ |
| $C_v$ | heat capacity at constant volume per unit mass | $J\,kg^{-1}K^{-1}$ |
| $C_D$ | drag coefficient | — |
| $d$ | diameter of particle | m |
| $\bar{d}$ | surface mean particle diameter | m |
| $d_t$ | diameter of tank, column or bed | m |
| $e$ | void fraction | — |
| $e_{mf}$ | void fraction at incipient fluidisation | — |
| $F$ | energy loss due to friction | $J\,kg^{-1}$ |
| $Fr_{mf}$ | Froude number, based on minimum fluidising velocity | — |
| $f_w$ | wall correction factor | — |
| $G$ | mass flow rate of fluid per unit cross-sectional area of bed or vessel | $kg\,m^{-2}s^{-1}$ |
| $Ga$ | Galileo number | — |
| $g$ | acceleration due to gravity | $m\,s^{-2}$ |
| $i$ | one species of a distribution | — |
| $K''$ | Kozeny constant | — |
| $L$ | depth of column or bed | m |

| Symbol | Description | Unit |
|---|---|---|
| $L_0$ | depth of packed bed at zero voidage | m |
| $L_{mf}$ | depth of packed bed at incipient fluidisation | m |
| $P$ | pressure | $N\,m^{-2}$ |
| $-\Delta P$ | pressure drop | $N\,m^{-2}$ |
| $Q_{mf}$ | volumetric flow rate at incipient fluidisation | $m^3\,s^{-1}$ |
| $R'$ | resistance per unit projected area of particle | $N\,m^{-2}$ |
| $Re$ | Reynolds number | — |
| $Re_1$ | Reynolds number as defined in example 11.1.2 | — |
| $Re_p$ | Reynolds number for particle settling in a fluid | — |
| $Re_{p,mf}$ | Reynolds number for a settling particle, based on flow rate at incipient fluidisation | — |
| $S$ | surface area per unit volume of particle or packing | $m^{-1}$ |
| $S_c$ | surface area of container per unit volume of bed | $m^{-1}$ |
| $U$ | superficial linear velocity of fluid | $m\,s^{-1}$ |
| $U_{mf}$ | superficial linear velocity at incipient fluidisation | $m\,s^{-1}$ |
| $U_t$ | terminal settling velocity of particle | $m\,s^{-1}$ |
| $V$ | volume of packed bed | $m^3$ |
| $v$ | specific volume of fluid | $m^3\,kg^{-1}$ |
| $-W_p$ | energy supplied to pump or blower, per unit mass of fluid | $J\,kg^{-1}$ |
| $-W_s$ | shaft work per unit mass of fluid | $J\,kg^{-1}$ |
| $x$ | weight fraction of particular solids in a mixture | — |
| $z$ | distance in vertical direction | m |
| $\alpha$ | kinetic energy correction factor | — |
| $\gamma$ | ratio of heat capacities, $C_p/C_v$ | — |
| $\eta$ | efficiency | — |
| $\mu$ | absolute viscosity of fluid | $kg\,m^{-1}\,s^{-1}$ |
| $\rho$ | density of fluid | $kg\,m^{-3}$ |
| $\rho_s$ | density of solid particle | $kg\,m^{-3}$ |

# 12 Pneumatic conveying

## 12.1 Worked Examples

### 12.1.1 Flow patterns in pneumatic conveying. The transition from fluidised bed flow to moving bed flow

*Problem*

The particles of a solid synthetic detergent are very nearly spherical and have a mean diameter of 0.650 mm. Their density is 518 kg m$^{-3}$, although when loosely packed their density appears to be 321 kg m$^{-3}$.

It is planned to move this detergent vertically, using a pneumatic conveying system, through a circular duct which has a diameter of 25.5 cm. If 22.7 kg of detergent are to be transported per minute, determine the linear air flow rate through the duct at which the transition from dense phase flow to moving bed flow should occur.

Density of air = 1.22 kg m$^{-3}$
Viscosity of air = 1.73 × 10$^{-5}$ kg m$^{-1}$ s$^{-1}$
Acceleration due to gravity = 9.81 m s$^{-2}$

*Solution*

In the design of a vertical pneumatic conveying system it is important to be able to predict, for different velocities, the flow regime encountered. For dense phase conveying the slip velocity between gas and solid is higher than that required for minimum fluidisation. The equation for predicting the transition from dense phase flow to moving bed flow is thus derived by equating the slip velocity in vertical pneumatic conveying with the slip velocity at incipient fluidisation (ref. 1, 2).

Thus, 
$$U^* - U_s^* = U_{mf}^* \tag{12.1}$$

At this transition point the voidage in the conduit equals the voidage at minimum fluidisation. Thus the actual linear velocities in equation (12.1) can be converted to superficial linear velocities by noting that

$$U^* = U/e_{mf}$$
$$U_s^* = U_s/(1-e_{mf})$$
$$U_{mf}^* = U_{mf}/e_{mf}$$

Substitution in equation (12.1) gives

$$\frac{U}{e_{mf}} - \frac{U_s}{1-e_{mf}} = \frac{U_{mf}}{e_{mf}}$$

or
$$(1-e_{mf})U - e_{mf}U_s = (1-e_{mf})U_{mf} \tag{12.2}$$

Furthermore, $e_{mf}$ may be determined from the density of the solids ($\rho_s$) and the mean bulk density ($\bar{\rho}$) from the equation

$$\bar{\rho} = (1-e_{mf})\rho_s + e_{mf}\rho$$

In the present problem, $\rho_s = 518 \text{ kg m}^{-3}$ and $\bar{\rho} = 321 \text{ kg m}^{-3}$, so that

$$321 = (1-e_{mf})518 + 1.22 e_{mf}$$

and therefore
$$e_{mf} = 0.381$$

The pressure drop in moving bed flow can be calculated from a modified Ergun-type equation (ref. 3 and 4), using the appropriate slip velocity,

i.e.
$$\frac{-\Delta P}{L} = 150 \frac{(1-e_{mf})^2}{(e_{mf})^3} \frac{(\mu U_{mf})}{d^2} + 1.75 \frac{(1-e_{mf})}{(e_{mf})^3} \frac{\rho(U_{mf})^2}{d}$$

When this equation is combined with the force balance equation for fluidised beds (see chapter 11) at incipient fluidisation, i.e.

$$\frac{-\Delta P}{L} = (1-e_{mf})(\rho_s - \rho)g$$

we obtain the quadratic equation in $U_{mf}$, namely

$$(1-e_{mf})(\rho_s - \rho)g = 150 \frac{(1-e_{mf})^2}{(e_{mf})^3} \frac{\mu U_{mf}}{d^2} + 1.75 \frac{(1-e_{mf})}{(e_{mf})^3} \frac{\rho(U_{mf})^2}{d}$$

All parameters in this equation are known except $U_{mf}$. Substitution gives

$$(1-0.381)(518-1.22)(9.81) = 150 \times \frac{(1-0.381)^2}{0.381^3} \times \frac{(1.73 \times 10^{-5})U_{mf}}{(0.650 \times 10^{-3})^2}$$

$$+ 1.75 \frac{(1-0.381)}{0.381^3} \frac{1.22 \, U_{mf}^2}{(0.650 \times 10^{-3})}$$

## 242 Problems in fluid flow

or
$$36\,760\,U_{mf}^2 + 42\,550\,U_{mf} - 3138 = 0$$

Solving this equation and taking only the positive root gives

$$U_{mf} = 0.070 \text{ m s}^{-1}$$

Now equation (12.2), after rearrangement, gives

$$U = e_{mf}\left(\frac{U_{mf}}{e_{mf}} + \frac{U_s}{1 - e_{mf}}\right) \qquad (12.3)$$

To determine $U_s$, the superficial linear velocity of the solids, we note that the mass flow rate of solids, $\dot{m}_s$, is equal to 22.7 kg min$^{-1}$,

i.e. $\dot{m}_s = 0.378 \text{ kg s}^{-1}$

The mass flux of solids, $G_s$, is given by

$$G_s = \frac{\dot{m}_s}{A}$$

$$= \frac{0.378}{\pi(12.75 \times 10^{-2})^2} \text{ kg m}^{-2}\text{s}^{-1}$$

$$= 7.40 \text{ kg m}^{-2}\text{s}^{-1}$$

Now the superficial linear velocity of the solid particles,

$$U_s = G_s/\rho_s$$
$$= 7.40/518 \text{ m s}^{-1}$$
$$= 1.43 \times 10^{-2} \text{ m s}^{-1}$$

Substitution of the calculated values of $e_{mf}$, $U_{mf}$ and $U_s$ into equation (12.3) finally gives the superficial linear velocity of air at the required transition point.

Thus
$$U = 0.381\left(\frac{0.070}{0.381} + \frac{1.43 \times 10^{-2}}{(1 - 0.381)}\right) \text{ m s}^{-1}$$

$$= 0.0788 \text{ m s}^{-1}$$

The actual linear air flow rate through the duct, $U^*$, is related to the superficial linear air velocity by

$$U^* = U/e_{mf}$$
$$= 0.0788/0.381 \text{ m s}^{-1}$$
$$= 0.21 \text{ m s}^{-1}$$

The transition from dense phase flow to moving bed flow will therefore occur at a linear air flow rate through the vertical duct of 0.21 m s$^{-1}$.

## 12.1.2 The prediction of choking velocity and choking voidage in a vertical transport line in which dilute phase flow occurs

*Problem*

Spherical polyethylene particles with a diameter of 3.4 mm are to be transported pneumatically through a vertical duct which has a diameter of 3.54 cm. A requirement is that the actual air velocity through the duct be no more than $15.0 \text{ m s}^{-1}$, in order to minimise energy requirements.
(i) Determine whether choking can occur in this system.
(ii) Estimate the maximum carrying capacity of the transport line, expressed as the mass flow rate of polyethylene particles.

Density of air = $1.22 \text{ kg m}^{-3}$
Density of polyethylene = $910 \text{ kg m}^{-3}$
Viscosity of air = $1.73 \times 10^{-5} \text{ kg m}^{-1} \text{ s}^{-1}$
Acceleration due to gravity = $9.81 \text{ m s}^{-2}$

*Solution*

(i) Choking does not occur in all gas-solid pneumatic conveying systems. The following criterion on whether choking will occur in a particular gas/solid/pipe system has been developed by Yang (ref. 5) and appears generally useful.

For no choking to occur

$$U_t/(g\, d_t)^{\frac{1}{2}} < 0.35$$

where $U_t$ is the terminal settling velocity of the particles which are to be pneumatically transported, and $d_t$ is the diameter of the transport line.

In the present case $U_t$ can be determined as outlined in Chapter 9, using the chart of $(R/\rho U_t^2)(Re_1)^2$ versus $Re_1$ which is given in reference 6.

Thus 
$$(R/\rho U_t^2)(Re_1)^2 = \frac{2d^3 \rho g (\rho_s - \rho)}{3\mu^2}$$

$$= \frac{2(3.4 \times 10^{-3})^3 \times 1.22 \times 9.81 \,(910 - 1.22)}{3(1.73 \times 10^{-5})^2}$$

$$= 9.522 \times 10^5$$

From the chart in reference 6,

$$Re_1 = 2.0 \times 10^3$$

Now 
$$Re_1 = \frac{d U_t \rho}{\mu}$$

## 244  Problems in fluid flow

So that
$$U_t = \frac{\mu Re_1}{d\rho}$$

$$= \frac{1.73 \times 10^{-5} \times 2.0 \times 10^3}{3.4 \times 10^{-3} \times 1.22} \text{ m s}^{-1}$$

$$= 8.34 \text{ m s}^{-1}$$

Therefore
$$U_t/(gd_t)^{\frac{1}{2}} = 8.34/(9.81 \times 3.54 \times 10^{-2})^{\frac{1}{2}}$$

$$= 14.2$$

This is greater than 0.35, so it is very likely that choking can occur in this particular gas/solid/pipe system.

(ii) For systems which have a choking transition the equations developed by Yang (ref. 2 and 5) were shown to correlate 90% of published data on choking point predictions to within ±30%. Yang's correlation, written in the following form, will be used here

$$U_{sc} = (U_c^* - U_t)(1 - e_c) \qquad (12.4)$$

$$(U_c^* - U_t)^2/2gd_t = 100(e_c^{-4.7} - 1) \qquad (12.5)$$

where $e_c$ is the voidage on choking, and $U_c^*$ the actual gas velocity at the choking point. Our method is to use equation (12.5) to calculate the voidage at choking, using the given superficial gas velocity, 15.0 m s$^{-1}$, as the superficial gas velocity at the choking point, and then to substitute $e_c$ into equation (12.4) to calculate $U_{sc}$, the superficial solids velocity.

Substitution of known values into equation (12.5) gives

$$\frac{(15.0 - 8.34)^2}{2 \times 9.81 \times 3.54 \times 10^{-2}} = 100(e_c^{-4.7} - 1)$$

whence
$$e_c = 0.900$$

This value of $e_c$ can now be substituted directly into equation (12.4) to give

$$U_{sc} = (15.0 - 8.34)(1 - 0.900) \text{ m s}^{-1}$$

$$= 0.666 \text{ m s}^{-1}$$

This is the superficial linear velocity of the polyethylene particles at the choking point. The superficial mass flux thus

$$= U_{sc}\rho_s$$
$$= 0.666 \times 910 \text{ kg m}^{-2}\text{ s}^{-1}$$
$$= 606 \text{ kg m}^{-2}\text{ s}^{-1}$$

Now the cross sectional area of the duct, $A$, is given by

$$A = \pi r^2$$
$$= \pi(3.54 \times 10^{-2}/2)^2 \text{ m}^2$$
$$= 9.84 \times 10^{-4} \text{ m}^2$$

It follows that the mass flow rate of polyethylene particles

$$= 606 \times 9.84 \times 10^{-4} \text{ kg s}^{-1}$$
$$= 0.60 \text{ kg s}^{-1}$$

The maximum carrying capacity of the transport line is therefore 0.60 kg of polyethylene particles per second.

## 12.1.3 The prediction of the pressure drop in dilute phase horizontal pneumatic transport

*Problem*

A sample of coal which is comprised of approximately spherical particles is to be transported pneumatically in a smooth horizontal circular duct which is 50 m long at a rate of 1.20 kg s$^{-1}$. The diameter of the coal particles is 0.75 mm, their density is 1400 kg m$^{-3}$, and laboratory measurements showed that their terminal settling velocity in air is 2.80 m s$^{-1}$.

If the mass ratio of solids to air which is to be used is 10:1, and if the superficial linear air velocity in the duct is to be 25.0 m s$^{-1}$, estimate the overall pressure drop along the duct.

Density of air = 1.22 kg m$^{-3}$
Viscosity of air = $1.73 \times 10^{-5}$ kg m$^{-1}$ s$^{-1}$
Acceleration due to gravity = 9.81 m s$^{-2}$

*Solution*

The pressure drop along the duct can be considered as the sum of three terms, namely (i) that due to the acceleration of the solid particles and air, (ii) that due to the force necessary to lift particles and air against gravity, and (iii) the pressure drop due to friction (ref. 1, 2, 7).

As the pipe is horizontal, item (ii) is negligible, and the pressure difference over a length $L$ of horizontal pipe can be written as (ref. 1, 2, 7).

$$-\Delta P = \Delta [\rho e (U^*)^2 + \rho_s (1-e) (U_s^*)^2] - \Delta P_f \qquad (12.6)$$

If it is assumed that the acceleration of both air and solids is done in the pipe,

246   Problems in fluid flow

then equation (12.6) simplifies to

$$-\Delta P = \rho e(U^*)^2 + \rho_s(1-e)(U_s^*)^2 - \Delta P_f \qquad (12.7)$$

To calculate the pressure drop $(-\Delta P)$ we must therefore estimate the actual gas velocity, $U^*$, the actual solids velocity, $U_s^*$, and the voidage, $e$, in addition to the pressure drop due to friction, $-\Delta P_f$.

To calculate the actual solids velocity, $U_s^*$, the superficial velocity, $U_s$, is required. Now the mass flow rate, $\dot{m}_s$, of coal is $1.20\ \text{kg s}^{-1}$. The volumetric flow rate of solids,

$$\begin{aligned}Q_s &= \dot{m}_s/\rho_s \\ &= 1.20/1400\ \text{m}^3\ \text{s}^{-1} \\ &= 8.57 \times 10^{-4}\ \text{m}^3\ \text{s}^{-1}\end{aligned}$$

To convert this volumetric flow rate to a superficial linear velocity, $U_s$, the cross-sectional area of the inside of the pipe is required, and this may be obtained by considering the mass ratio of solids to air and the superficial linear air velocity, both of which are given in the problem statement.

Thus the mass flow rate of solids $= 1.2\ \text{kg s}^{-1}$ so the mass flow rate of air $= 1.2/10 = 0.12\ \text{kg s}^{-1}$.

The volumetric flow rate of air, $Q_g$, can then be found as

$$\begin{aligned}Q_g &= \dot{m}_g/\rho \\ &= 0.12/1.22\ \text{m}^3\ \text{s}^{-1} \\ &= 0.0984\ \text{m}^3\ \text{s}^{-1}\end{aligned}$$

Now as the superficial linear velocity of the air, $U$, is related to $Q_g$ by

$$U = Q_g/A,$$

then

$$\begin{aligned}A &= Q_g/U \\ &= 0.0984/25.0 = 3.936 \times 10^{-3}\ \text{m}^2\end{aligned}$$

Referring back to the solids flow rate, as

$$Q_s = 8.57 \times 10^{-4}\ \text{m}^3\ \text{s}^{-1},$$

then

$$\begin{aligned}U_s &= Q_s/A \\ &= (8.57 \times 10^{-4})/(3.936 \times 10^{-3})\ \text{m s}^{-1} \\ &= 0.218\ \text{m s}^{-1}\end{aligned}$$

Now the actual linear velocity, $U_s^*$, of the solids may be related to the superficial linear gas rate $(U)$ by the equation (ref. 8),

$$U_t = U - U_s^*$$

where $U_t$ is the terminal settling velocity of the coal particles, which is given, and thus

$$\begin{aligned}U_s^* &= U - U_t \\ &= 25.0 - 2.80 = 22.20\ \text{m s}^{-1}\end{aligned}$$

To determine the voidage, $e$, the following relationships may now be used.

$$U_s^* = U_s/(1-e)$$

or

$$e = \frac{U_s^* - U_s}{U_s^*}$$

$$= \frac{22.20 - 0.218}{22.20} = 0.9902$$

We now have all the information required to calculate the first two terms on the right-hand side of equation (12.7).

The third term, the pressure drop due to friction, $-\Delta P_f$, is usually taken to be the sum of two terms, (i) the pressure drop due to friction in the air alone $(-\Delta P_a)$, and (ii) the friction due to the presence of the solids, $(-\Delta P_s)$.

Thus

$$\Delta P_f = \Delta P_a + \Delta P_s \tag{12.8}$$

Now $-\Delta P_a$ can be evaluated using a usual equation for fluid friction (ref. 9, 10), i.e.

$$F = \frac{2\phi U^2 L}{r}$$

where $F$ is the energy loss due to fluid friction per unit mass of fluid, so that

$$-\Delta P_a = \rho F$$

$$= \frac{2\phi U^2 L \rho}{r} \tag{12.9}$$

To determine $\phi$, the friction factor, the Reynolds number for the air flowing through the duct must be known.

Now

$$Re = \frac{d_t U \rho}{\mu}$$

where

$$d_t = \sqrt{4A/\pi}$$

$$= \sqrt{4 \times 3.936 \times 10^{-3}/\pi} \text{ m}$$

$$= 0.0708 \text{ m}$$

so that

$$Re = \frac{0.0708 \times 1.22 \times 25.0}{1.73 \times 10^{-5}}$$

$$= 1.25 \times 10^5$$

Using the friction factor chart (ref. 11, 12) we find $\phi = 2.1 \times 10^{-3}$.
Therefore, from equation (12.9)

$$-\Delta P_a = \frac{2 \times 2.1 \times 10^{-3} \times 25.0^2 \times 50 \times 1.22}{0.0708/2} \text{ N m}^{-2}$$

$$= 4523 \text{ N m}^{-2}$$

248  Problems in fluid flow

To calculate the pressure drop $(-\Delta P_s)$ attributable to the solids alone, several methods are available. We shall use initially the solid particle friction factor $(f_s)$ as used by Leung and Wiles (ref. 1, 4).

Thus
$$-\Delta P_s = \frac{2Lf_s(1-e)\rho_s(U_s^*)^2}{d_t} \qquad (12.10)$$

where $f_s$ may be determined, also from Leung and Wiles work, as

$$f_s = \frac{0.05}{U_s^*}$$

Note that $U_s^*$ is the actual linear velocity of solids in the pipe, and must be in $\mathrm{m\,s^{-1}}$ for this equation, which is dimensional.
Thus for the present problem

$$f_s = \frac{0.05}{22.20} = 2.25 \times 10^{-3}$$

Substitution into equation (12.10) gives

$$-\Delta P_s = \frac{2(50)(2.25 \times 10^{-3})(1-0.9902)(1400)(22.20)^2}{0.0708} \mathrm{\,N\,m^{-2}}$$

$$= 21\,500 \mathrm{\,N\,m^{-2}}$$

All the terms in equation (12.7) can now be evaluated, and thus

$$-\Delta P = 1.22 \times 0.9902 \times 25.0^2 + 1400(1-0.9902)(22.20)^2 + 4523$$
$$+ 21\,500 \mathrm{\,N\,m^{-2}}$$

$$= 33\,500 \mathrm{\,Nm^{-2}}$$

The overall pressure drop along the pipe line is thus, approximately equal to $\underline{33\,500 \mathrm{\,N\,m^{-2}}}$.

*Alternative Estimation of* $-\Delta P_f$
The above solution makes use of the concept of the solid particle friction factor, as used by Leung and Wiles (ref. 1, 4). It is of interest to compare the calculated value of $-\Delta P_s$ with those obtained using other procedures.
For example, using the dimensional equation of reference 13,

$$U - U_s^* = \frac{U_t}{0.48 + 7.25\sqrt{U_t/\rho_s}}$$

$$= \frac{2.80}{0.48 + 7.25\sqrt{2.80/1400}} \mathrm{\,m\,s^{-1}}$$

$$= 3.48 \mathrm{\,m\,s^{-1}}$$

Pneumatic conveying 249

Thus
$$U_s^* = 25.0 - 3.48 \text{ m s}^{-1}$$
$$= 21.52 \text{ m s}^{-1}$$

Now $-\Delta P_a$, the pressure drop attributable to air, is, as calculated previously, 4523 N m$^{-2}$. Using the dimensional relationship given in reference 13,

$$\frac{-\Delta P_s}{-\Delta P_a} \frac{(U_s^*)^2}{\dot{m}_s} = \frac{2805}{U_t}$$

whence
$$-\Delta P_s = \frac{2805 \times 4523 \times 1.2}{2.80 \times 21.52^2} \text{ N m}^{-2}$$
$$= \underline{11\,700 \text{ N m}^{-2}}$$

Two other methods of estimating $-\Delta P_s$ are given in references 14 and 15. In the first of these

$$-\Delta P_f = 2f_s' \frac{U}{U_s^*} \frac{G_s}{G_a} \frac{\rho U^2 L}{d_t}$$

$$= 2f_s' \frac{\bar{\rho} U^2 L}{d_t}$$

From Table 6 of reference 15, $f_s'$ is approximately 0.0011.

Also
$$\bar{\rho} = \rho_s (1 - e) + \rho e$$
$$= 1400 (1 - 0.9902) + 1.22 \times 0.9902$$
$$= 14.9 \text{ kg m}^{-3}$$

Thus
$$-\Delta P_f = \frac{2 \times 0.0011 \times 14.9 \times 25.0^2 \times 50}{0.0708} \text{ N m}^{-2}$$
$$= \underline{14\,470 \text{ N m}^{-2}}$$

Using the second method given in reference 15, the value of $-\Delta P_a$ equals 4523 N m$^{-2}$ as before, so that

$$-\Delta P_s = \frac{\pi}{8} \cdot \frac{f_p}{f_g} \cdot \left(\frac{\rho_s}{\rho}\right)^{\frac{1}{2}} \left(\frac{G_s}{G_a}\right)(-\Delta P_a)$$

$$= \frac{\pi}{8} \frac{(0.7 \times 10^{-4})}{0.0042} \left(\frac{1400}{1.22}\right)^{\frac{1}{2}} 10 \times 4523 \text{ N m}^{-2}$$

$$= \underline{10\,030 \text{ N m}^{-2}}$$

Note that the values of the two friction factors, $f_g$ and $f_p$, have been determined using information provided in references 15 and 16. Thus the equation

$$f_g = 0.0791 \left(\frac{\rho U d_t}{\mu}\right)^{-0.25}$$

250  *Problems in fluid flow*

has been used to determine $f_g$, and the Figure 19 (ref. 16) for $f_p$.

## 12.1.4 The prediction of the pressure drop in dilute phase vertical pneumatic transport

*Problem*

Rape seed is to be transported pneumatically from a cargo ship to a silo through 18.5 m of vertical steel tube (absolute roughness = 0.046 mm) which has an inside diameter of 14.5 cm. A flow rate of 5 tonnes of rape seed per hour is required. The seed has a mean particle size of 1.78 mm and a density of 1090 kg m$^{-3}$. Laboratory measurements on the seed showed that its terminal settling velocity in air, under the conditions prevailing in the steel tube, would be 6.50 m s$^{-1}$.

A blower which is capable of supplying air at a rate of 0.40 m$^3$ s$^{-1}$ is available. What is the pressure drop along the length of the tube?

Density of air = 1.22 kg m$^{-3}$
Viscosity of air = 1.73 × 10$^{-5}$ kg m$^{-1}$ s$^{-1}$
Acceleration due to gravity = 9.81 m s$^{-2}$

*Solution*

As was the case for the previous problem, we shall consider the pressure drop to have three components, namely (i) that due to acceleration of solid particles and air, (ii) that required to raise solid particles and air against gravity and (iii) the pressure drop due to friction.

In this case the pressure drop over a length, $L$, of vertical pipe can be written in the following general form (ref. 1, 2 and 7):

$$-\Delta P = \Delta \left[\rho e \, (U^*)^2 + \rho_s \, (1-e) \, (U_s^*)^2\right] + g \int_0^L \left[(1-e)\rho_s + e\rho\right] dz - \Delta P_f \quad (12.11)$$

where the three terms on the right-hand side of equation (12.11) represent the pressure drop due to the acceleration of air and solids, raising the potential energy, and in overcoming friction. If it is assumed that the pressure drop in the vertical conduit is small compared with the absolute pressure, and that the acceleration section is short compared with the conduit length, and that the gas density is negligible, then equation (12.11) can be simplified to

$$-\Delta P = \rho_s(1-e)(U_s^*)^2 + (1-e)g\rho_s L - \Delta P_f \quad (12.12)$$

## Pneumatic conveying

To calculate the pressure drop we therefore need to estimate the actual solids velocity, $U_s^*$, the voidage, $e$, and the pressure drop due to friction, $-\Delta P_f$.

To calculate the actual solids velocity, $U_s^*$, the superficial solids velocity is first required. Now the mass flow rate of the rape seed, $\dot{m}_s$, is given by

$$\dot{m}_s = 5 \text{ tonnes per hour}$$
$$= 5 \times 10^3/3600 = 1.389 \text{ kg s}^{-1}$$

The volumetric flow rate of solids, $Q_s$, is given by

$$Q_s = \dot{m}_s/\rho_s$$
$$= 1.389/1090 = 1.274 \times 10^{-3} \text{ m}^3 \text{ s}^{-1}$$

The superficial solids flow rate, $U_s$, is given by
$$U_s = Q_s/A$$
$$= 1.274 \times 10^{-3}/[\pi(7.25 \times 10^{-2})^2] \text{ m s}^{-1}$$
$$= 7.715 \times 10^{-2} \text{ m s}^{-1}$$

Now the actual linear velocity ($U_s^*$) of the rape seed can be calculated from the equation

$$U_s^* = U_s/(1-e) \tag{12.13}$$

provided that voidage, $e$, at the operating conditions, can be calculated. If the superficial velocity of the gas ($U$), and the terminal settling velocity of the solid, $U_t$, are known then as

$$U_t = U - U_s^* \tag{12.14}$$

equations (12.13) and (12.14) together yield the actual velocity of the solids, $U_s^*$, and the voidage, $e$. This procedure assumes that the slip velocity equals the terminal settling velocity ($U_t$) of the solid particles, and that the superficial gas velocity ($U$) and the actual linear gas velocity ($U^*$) are also equal. As

$$U^* = U/e$$

and $e$ is expected to be greater than 0.99 in dilute phase pneumatic transport, the latter assumption is valid.

To use equation (12.14) to calculate the actual linear velocity ($U_s^*$) of the solids, the superficial linear gas velocity, $U$, is required.

The volumetric flow rate of the gas, $Q_g$, equals $0.40 \text{ m}^3\text{s}^{-1}$. The superficial gas velocity is given by

$$U = Q_g/A$$
$$= \frac{0.40}{\pi(7.25 \times 10^{-2})^2} = 24.22 \text{ m s}^{-1}$$

252  *Problems in fluid flow*

From equation (12.14)

$$U_s^* = U - U_t$$
$$= 24.22 - 6.50 = 17.72 \text{ m s}^{-1}$$

Substitution of the now known values of $U_s$ and $U_s^*$ in equation (12.13) gives

$$17.72 = \frac{7.715 \times 10^{-2}}{1-e}$$

whence
$$e = 0.9956$$

The first two terms of the right-hand side of equation (12.12) can now be calculated. The third term, the pressure drop due to friction, $-\Delta P_f$, is as before considered to be the sum of two terms, namely the pressure drop due to fluid friction, $-\Delta P_a$, and the pressure drop attributable to the presence of the solids, $-\Delta P_s$.

Thus
$$\Delta P_f = \Delta P_a + \Delta P_s \qquad (12.15)$$

The first term on the right-hand side of equation (12.15) can be evaluated using the equation

$$F = \frac{2\phi U^2 L}{r}$$

where $F$ is the energy loss due to fluid friction (ref. 9, 10) in $\text{J kg}^{-1}$, so that as

$$\frac{-\Delta P_a}{\rho} = F$$

it follows that

$$-\Delta P_a = \frac{2\phi U^2 L \rho}{r} \qquad (12.16)$$

To determine the friction factor, $\phi$, the Reynolds number, $Re$, must be known.

$$Re = \frac{d_t \rho U}{\mu}$$

$$= \frac{14.5 \times 10^{-2} \times 1.22 \times 24.22}{1.73 \times 10^{-5}} = 2.47 \times 10^5$$

Since the absolute roughness of the pipe = 0.046 mm, the relative roughness of the pipe

$$= \frac{0.046 \times 10^{-3}}{14.5 \times 10^{-2}} = 3.17 \times 10^{-4}$$

Using this value of the pipe roughness, and the calculated value above for the Reynolds number due to gas flow only, the friction factor, $\phi$, can be read from the chart of $\phi$ versus $Re$ (Appendix, Fig. A1)

Thus
$$\phi = 2.08 \times 10^{-3}$$

Substitution of known values in equation (12.16) now gives

$$-\Delta P_a = \frac{2 \times 2.08 \times 10^{-3} \times 24.22^2 \times 18.5 \times 1.22}{7.25 \times 10^{-2}} \, \text{N m}^{-2}$$

$$= 760 \, \text{N m}^{-2}$$

To calculate the second term ($\Delta P_s$) on the right-hand side of equation (12.15), use will be made of the solid particle friction factor ($f_s$) referred to in example 12.1.3.

Thus
$$-\Delta P_s = \frac{2L f_s (1-e)\rho_s (U_s^*)^2}{d_t} \tag{12.17}$$

where $f_s$ is perhaps best determined from the simple equation (ref. 1 and 4)

$$f_s = \frac{0.05}{U_s^*}$$

($U_s^*$ must be in m s$^{-1}$ in this equation)
Thus in the present situation,

$$f_s = \frac{0.05}{17.72} = 2.82 \times 10^{-3}$$

and so from equation (12.17) we get

$$-\Delta P_s = \frac{2 \times 18.5 \times 2.82 \times 10^{-3} \times (1-0.9956) \times 1090 \times 17.72^2}{14.5 \times 10^{-2}} \, \text{N m}^{-2}$$

$$= 1084 \, \text{N m}^{-2}$$

Hence from equation (12.15) the total pressure drop, $-\Delta P_f$, due to the two friction contributions is given by

$$-\Delta P_f = 1084 + 760 = 1844 \, \text{N m}^{-2}$$

We are now in a position to make all the substitutions required in equation (12.12) in order to calculate the overall pressure drop. Thus

$$-\Delta P = (1090 \times (1-0.9956) \times 17.72^2) + [(1-0.9956) \times 9.81 \times 1090 \times 18.5] + 1844 \, \text{N m}^{-2}$$

$$= 4220 \, \text{N m}^{-2}$$

The overall pressure drop in the pipe is thus approximately 4220 N m$^{-2}$.

## 12.1.5 The dense phase flow regime for the pneumatic transport of solids, upwards through vertical tubes without slugging

*Problem*

An attempt is to be made to transport an ore sample upwards through a vertical pipe which is 25.0 metres long. The flow pattern obtained may be assumed to be in the dense flow regime of pneumatic transport, but without slugging.

Estimate the pressure drop which may be expected along the pipe.

Density of ore = 2650 kg m$^{-3}$
Acceleration due to gravity = 9.81 m s$^{-2}$

*Solution*

In the dense phase flow regime without slugging the frictional pressure drop is usually small compared with that due to the mass of the solids being transported. Thus overall pressure drop,

$$-\Delta P = \rho_s (1-e) g L \quad (12.18)$$

To calculate $-\Delta P$, the value of the voidage, $e$, for the system is required. Unfortunately no reliable equations for estimating the bed voidage are available. The voidage can be determined experimentally in a test rig, or as a first approximation it may be assumed to be 0.6–0.8, depending on the operating velocity (ref. 1).

Assuming a voidage of 0.6, equation (12.18) becomes,

$$-\Delta P = 2650 (1-0.6)\, 9.81 \times 25.0 \text{ N m}^{-2}$$
$$= 2.6 \times 10^5 \text{ N m}^{-2}$$

If a voidage of 0.8 is assumed,

$$-\Delta P = 2650 (1-0.8)\, 9.81 \times 25.0 \text{ N m}^{-2}$$
$$= 1.3 \times 10^5 \text{ N m}^{-2}$$

Thus the pressure drop in the transport line is probably between $1.3 \times 10^5$ N m$^{-2}$ and $2.6 \times 10^5$ N m$^{-2}$. Note however that these calculations assume that the pressure drop along the line is small compared with the absolute pressure at any point in the system. This is unlikely to be the case, as the calculated values of $-\Delta P$ are both comparable to atmospheric pressure (approximately $1.01 \times 10^5$ N m$^{-2}$) which is most likely to be the pressure obtaining at the top end of the transport line. Experimental data are therefore desirable for accurate design work.

## 12.2 Student Exercises

1  Determine whether dense phase slugging flow is likely to occur when the following materials are pneumatically transported at very high solids to air ratios through a vertical pipe which has an inside diameter of 1.0 cm. The air has a temperature of 37.8 °C and a pressure of 101.3 kPa.
(i) polystyrene beads of diameter 0.05 mm, and density 1080 kg m$^{-3}$
(ii) spherical coal particles of density 1480 kg m$^{-3}$ which have a terminal settling velocity of 2.80 m s$^{-1}$ in the air.

Density of air at 37.8 °C and 101.3 kPa = 1.137 kg m$^{-3}$
Viscosity of air at 37.8 °C and 101.3 kPa = 1.90 × 10$^{-5}$ kg m$^{-1}$ s$^{-1}$
Acceleration due to gravity = 9.81 m s$^{-2}$

2  Spheres of polystyrene foam with a diameter of 5.0 mm and a density of 16.23 kg m$^{-3}$ have a void fraction of 0.4 when loosely packed. The spheres have to be transported pneumatically through a vertical tube which has an inside diameter of 7.02 cm and a length of 12.2 m at a rate of 25 kg per minute.
What is the minimum air velocity that is required to ensure that moving bed flow does not occur?

Density of air = 1.218 kg m$^{-3}$
Viscosity of air = 1.73 × 10$^{-5}$ kg m$^{-1}$ s$^{-1}$
Acceleration due to gravity = 9.81 m s$^{-2}$

3  Estimate the volumetric air flow rate required for the pneumatic conveyance of micro-spheroidal cracking catalyst particles, of diameter 0.150 mm and density 1610 kg m$^{-3}$, through a horizontal pipeline which has a diameter of 5.12 cm. A flow rate of 0.48 kg of catalyst per second is required, and the particles have a terminal settling velocity of 0.699 m s$^{-1}$ in air.

Density of air = 1.22 kg m$^{-3}$
Viscosity of air = 1.75 × 10$^{-5}$ kg m$^{-1}$ s$^{-1}$
Acceleration due to gravity = 9.81 m s$^{-2}$

4  A granulated fertiliser which has a density of 2503 kg m$^{-3}$ and a particle diameter of 5.0 × 10$^{-6}$ m is to be removed from a bin at a rate of 2.0 tonnes per hour. A pneumatic conveying line is to be used, which comprises a smooth vertical pipe with an inside diameter of 18.5 cm. The mass ratio of solids to air in the system is to be 5.0. Assuming fully developed flow, determine the following:
(i) the pressure drop per metre length of pipe which is due to friction attributable to the air,
(ii) the pressure drop per metre length of pipe which is attributable to the solids,

256  Problems in fluid flow

(iii) the total pressure drop, per metre length of pipe.

Density of air = 1.22 kg m$^{-3}$
Viscosity of air = 1.73 × 10$^{-5}$ kg m$^{-1}$ s$^{-1}$
Acceleration due to gravity = 9.81 m s$^{-2}$

5   Sand has to be moved at a rate of 3 tonnes per hour through a smooth, circular, horizontal duct which has an inside diameter of 15.75 cm and includes two right-angle bends. The sand particles are spherical, with a diameter of 0.12 mm, a density of 2655 kg m$^{-3}$, and a terminal settling velocity in air of 1.10 m s$^{-1}$. It is recommended that the actual linear velocity of the particles along the duct be no greater than 3.5 m s$^{-1}$ to prevent damage to the system.

The blower available can provide air at 14.0 m$^3$ per minute, but the pressure developed at the entrance to the duct is limited to 1.030 × 10$^5$ N m$^{-2}$ absolute.
(i) What is the pressure drop in the system which is attributable to the two right-angle bends?
(ii) In view of the pressure restriction, what is the maximum length of duct that can be used?
(iii) What would be the maximum possible length for the duct if the two bends could be eliminated and the duct was straight?

Density of air = 1.22 kg m$^{-3}$
Viscosity of air = 1.73 × 10$^{-5}$ kg m$^{-1}$ s$^{-1}$
Acceleration due to gravity = 9.81 m s$^{-2}$
Atmospheric pressure = 1.013 × 10$^5$ N m$^{-2}$

6   Perspex particles have to be transported in air through a horizontal pipe of inside diameter 10.4 cm at a rate of 0.33 kg s$^{-1}$. If the density of the Perspex is 1200 kg m$^{-3}$, and the terminal settling velocity of the particles in air is 4.85 m s$^{-1}$, determine the saltation velocity for the system.

Acceleration due to gravity = 9.81 m s$^{-2}$

## 12.3   References

Five references are abbreviated as follows:

C and R (1) = J. M. Coulson and J. F. Richardson, Chemical Engineering, Vol. 1, 3rd edition, Pergamon, 1977

C and R (2) = J. M. Coulson and J. F. Richardson, Chemical Engineering, Vol. 2, 3rd edition, Pergamon, 1980

F et al. = A. S. Foust, L. A. Wenzel, C. W. Clump, L. Maus and L. B. Andersen, Principles of Unit Operations, 2nd edition, John Wiley and Sons, 1980

K and L = D. Kunii and O. Levenspiel, Fluidisation Engineering, John Wiley and Sons, 1969

McC, S & H = W. L. McCabe, J. C. Smith and P. Harriott, *Unit Operations of Chemical Engineering*, 4th edition, McGraw-Hill, 1985

1. L. S. Leung and R. J. Wiles, *Ind. Eng. Chem., Process Des. Dev.*, 1976, **15**, 552–7
2. F et al., page 647
3. L. S. Leung and R. J. Wiles, *Ind. Eng. Chem., Process Des. Dev.*, 1976, **15**, 555
4. F et al., page 648
5. W. Yang, *A. I. Ch. E. Journal*, 1975 **21**, 1013–15
6. C and R (2), page 98
7. S. Stermerding, *Chem. Eng. Sc.*, 1962 **17**, 599–608
8. F et al., page 650
9. C and R (1), page 43
10. McC, S & H, chapter 5
11. C and R (1), page 42
12. McC, S & H, page 88
13. C and R (2), pages 285 and 286
14. K and L, page 389
15. K and L, page 390
16. K and L, page 391

## 12.4 Notation

| Symbol | Description | Unit |
|---|---|---|
| $A$ | cross sectional area of duct | $m^2$ |
| $d$ | particle diameter | m |
| $d_t$ | diameter of tank, duct or pipe | m |
| $e$ | voidage of air-solids mixtures | — |
| $e_c$ | voidage at choking point | — |
| $e_{mf}$ | voidage at incipient fluidisation | — |
| $F$ | energy loss due to friction | $J\,kg^{-1}$ |
| $f_g$ | gas friction factor (in ref. 15) | — |
| $f_p$ | friction factor (in ref. 15) | — |
| $f_s$ | solid particle friction factor (in ref. 1 and 4) | — |
| $f'_s$ | solid particle friction factor (in ref. 14) | — |
| $g$ | acceleration due to gravity | $m\,s^{-2}$ |
| $G_a$ | mass flux of air | $kg\,m^{-2}s^{-1}$ |
| $G_s$ | mass flux of solids | $kg\,m^{-2}s^{-1}$ |
| $L$ | length of pipe or duct | m |
| $\dot{m}_g$ | mass flow rate of air | $kg\,s^{-1}$ |
| $\dot{m}_s$ | mass flow rate of solids | $kg\,s^{-1}$ |
| $-\Delta P$ | total or overall pressure drop | $N\,m^{-2}$ |
| $-\Delta P_a$ | pressure drop attributable to air only | $N\,m^{-2}$ |
| $-\Delta P_f$ | pressure drop due to total friction loss | $N\,m^{-2}$ |
| $-\Delta P_s$ | pressure drop attributable to solids only | $N\,m^{-2}$ |
| $Q_g$ | volumetric flow rate of air | $m^3\,s^{-1}$ |
| $Q_s$ | volumetric flow rate of solids | $m^3\,s^{-1}$ |
| $r$ | radius of pipe or tank | m |
| $R$ | wall shear stress | $N\,m^{-2}$ |
| $Re$ | Reynolds number of the air | — |
| $Re_1$ | Reynolds number as defined in example 12.1.2 | — |
| $U$ | superficial air velocity | $m\,s^{-1}$ |
| $U^*$ | actual air velocity | $m\,s^{-1}$ |
| $U_c$ | superficial air velocity at choking point | $m\,s^{-1}$ |

| Symbol | Description | Unit |
|---|---|---|
| $U_c^*$ | actual air velocity at choking point | $\text{m s}^{-1}$ |
| $U_{mf}$ | superficial air velocity at incipient fluidisation | $\text{m s}^{-1}$ |
| $U_{mf}^*$ | actual air velocity at incipient fluidisation | $\text{m s}^{-1}$ |
| $U_s$ | superficial solids velocity | $\text{m s}^{-1}$ |
| $U_s^*$ | actual solids velocity | $\text{m s}^{-1}$ |
| $U_{sc}$ | superficial solids velocity at choking point | $\text{m s}^{-1}$ |
| $U_{sc}^*$ | actual solids velocity at choking point | $\text{m s}^{-1}$ |
| $U_t$ | terminal settling velocity of particle | $\text{m s}^{-1}$ |
| $\Delta U$ | linear velocity of air relative to solids | $\text{m s}^{-1}$ |
| $\mu$ | viscosity of air | $\text{kg m}^{-1}\text{s}^{-1}$ |
| $\bar{\rho}$ | mean bulk density of mixture | $\text{kg m}^{-3}$ |
| $\rho$ | density of air | $\text{kg m}^{-3}$ |
| $\rho_s$ | density of solids | $\text{kg m}^{-3}$ |
| $\phi$ | gas phase friction factor | — |

# 13 Centrifugal separation operations

## 13.1 Worked Examples

### 13.1.1 Equations for centrifugal force

*Problem*

(i) A laboratory centrifuge which has a bowl with a radius of 2.55 cm and a depth of 10.25 cm rotates at a speed of 1055 revolutions per minute. When used to clarify an aqueous solution the overflow weir is set to give a liquid depth of 1.51 cm on the surface of the bowl. Calculate the centrifugal force developed by the centrifuge. Express the answer in terms of gravity forces.
(ii) Calculate the cross-sectional area of a gravity settling tank which has the same clarifying capacity as the centrifuge system.

Density of water = 998 kg m$^{-3}$
Acceleration due to gravity = 9.81 m s$^{-2}$

*Solution*

(i) The acceleration of a particle in a centrifugal field is given by

$$a = r\omega^2$$

where $r$ is the radial distance of the particle from the centre of rotation and $\omega$ is its angular velocity in radians per second. Thus the force ($F_c$) acting on a particle in a centrifugal field is given by

$$F_c = ma$$
$$= mr\omega^2$$

Furthermore, the force ($F_g$) acting on the same particle in a gravitational field would be given by

$$F_g = mg$$

It follows that
$$\frac{F_c}{F_g} = \frac{r\omega^2}{g}$$

Thus the centrifugal force developed in a centrifuge is $r\omega^2/g$ times the force due to gravity.

For the present problem $r = 0.0255$ m and $g = 9.81$ m s$^{-2}$.

$$\omega = 2\pi N/60$$
$$= 2\pi \times 1055/60 \text{ rad s}^{-1}$$
$$= 110.5 \text{ rad s}^{-1}$$

Thus $\quad F_c/F_g = 0.0255 \times 110.5^2/9.81$
$$= \underline{31.74}$$

The centrifuge thus develops a force of 31.7 times the force due to gravity.

(ii) From the elementary theory of sedimentation in a centrifugal field (ref. 1, 2 and 3) it can be shown that the volumetric feed rate, $Q$, of a suspension into a centrifuge is given by

$$Q = \frac{d^2(\rho_s - \rho)g}{18\mu} \cdot \frac{\omega^2 V}{g \ln(R/r_i)} \tag{13.1}$$

Now providing the rate of settling of particles within the centrifuge falls in the Stokes law range, then the terminal settling velocity, $U_t$, of a particle is given by

$$U_t = d^2(\rho_s - \rho)g/18\mu \tag{13.2}$$

By combining equations (13.1) and (13.2) it follows that

$$Q/U_t = \omega^2 V/[g \ln(R/r_i)] \tag{13.3}$$

Since $Q$ is the volumetric feed rate, and $U_t$ the settling velocity, $Q/U_t$ is the cross-sectional area required for a tank in which the settling achieved is the same as that achieved in the centrifuge. Denoting this area by $\Sigma$,

$$\Sigma = \omega^2 V/[g \ln(R/r_i)] \tag{13.4}$$

Now $V$, the volumetric capacity of the centrifuge, is given by

$$V = \pi(R^2 - r_i^2)L$$
$$= \pi(0.0255^2 - 0.0104^2)(0.1025) \text{ m}^3$$
$$= 1.746 \times 10^{-4} \text{ m}^3$$

(Note that $r_i = R - h = 0.0255 - 0.0151 = 0.0104$ m)

The substitution of this and other known values into equation (13.4) gives

$$\Sigma = (110.5^2 \times 1.746 \times 10^{-4})/[9.81 \ln(0.0255/0.0104)] \text{ m}^2$$
$$= \underline{0.242 \text{ m}^2}$$

The centrifuge is thus equivalent to a gravity settling tank with a cross-sectional area of 0.242 m$^2$.

## 13.1.2 Fluid pressure in a tubular-bowl centrifuge. Maximum safe speed of rotation

*Problem*

A centrifuge has a bronze bowl which is 35.2 cm in diameter and 4.325 mm thick. When the centrifuge is used to clarify a solution of density 1105 kg m$^{-3}$, a layer of liquid which is 80 mm thick forms on the wall of the bowl.

The safe working stress of bronze may be taken as 55.0 MN m$^{-2}$, and its density is 8890 kg m$^{-3}$.
(i) What is the maximum safe speed of rotation for the centrifuge in this operation?
(ii) What is the centrifugal pressure due to the liquid on the wall of the bowl?
(iii) What is the pressure gradient in the liquid adjacent to the wall of the bowl?

*Solution*

(i) A force balance in the radial direction (refs. 1, 4 and 5) leads to the following equation for the centrifugal pressure at the wall of a centrifuge basket of radius $R$

$$P = \tfrac{1}{2}\rho\omega^2(R^2 - r_i^2) \qquad (13.5)$$

Now for the present problem $R = 0.352/2 = 0.176$ m

and
$$r_i = R - h$$
$$= 0.176 - 0.080 = 0.096 \text{ m}$$

It follows, from equation (13.5), that

$$P = 0.5 \times 1105 \times \omega^2(0.176^2 - 0.096^2) \text{ N m}^{-2}$$
$$= 12.02\,\omega^2 \text{ N m}^{-2} \qquad (13.6)$$

Now the stress on the walls of the centrifuge basket is given by (ref. 6),

$$S = (R/\tau)(P + \rho_b \tau R \omega^2)$$

Thus

$$55.0 \times 10^6 = (0.176/4.325 \times 10^{-3})$$
$$\times [12.02\omega^2 + (8890 \times 4.325 \times 10^{-3} \times 0.176\omega^2)]$$
$$= 764.5\,\omega^2 \text{ N m}^{-2}$$
$$\omega = \sqrt{55.0 \times 10^6/764.5} \text{ rad s}^{-1}$$
$$= 268.2 \text{ rad s}^{-1}$$

Now $\qquad \omega = 2\pi N/60$
Thus $\qquad N = 60\omega/2\pi$
$\qquad\qquad = \underline{2560}$ revolutions per minute.

262  Problems in fluid flow

This therefore is the maximum safe speed of rotation.

(ii) The centrifugal pressure, P, of the liquid at the wall of the centrifuge bowl can now be obtained from equation (13.6) since $\omega$ is known.

Thus $\qquad \omega = 268.2 \text{ rad s}^{-1}$,

and $\qquad P = 12.02\omega^2 \text{ N m}^{-2}$

$\qquad\qquad = 12.02 \times 268.2^2 \text{ N m}^{-2}$

$\qquad\qquad = \underline{8.65 \times 10^5 \text{ N m}^{-2}}$

(iii) In the derivation of equation (13.5) (ref. 1 and 5) the following expression for the pressure gradient in the liquid layer within a centrifuge, in the radial direction, is obtained,

$$\frac{dP}{dr} = \rho\omega^2 r$$

Now at the wall, $r = R$, and the pressure gradient in the radial direction

$\qquad = \rho\omega^2 R$

$\qquad = 1105 \times 268.2^2 \times 0.176 \text{ N m}^{-2}$ per metre

$\qquad = \underline{1.40 \times 10^7 \text{ N m}^{-2}}$ per metre

### 13.1.3 Particle size determination of fine particles by batch centrifugation

*Problem*

A finely divided sample of a water insoluble organic pigment of density 1425 kg m$^{-3}$ was thoroughly dispersed in water and the suspension transferred to a centrifuge tube. The surface of the liquid in the tube was 16.50 cm from the axis of rotation of the centrifuge.

After centrifuging for 360 s at 360 revolutions per minute a pipette was lowered into the centrifuge tube and a small sample of the suspension withdrawn from 1.00 cm below the surface.

Determine the particle size of the organic pigment which was suspended in this sample.

Density of water = 998 kg m$^{-3}$

Viscosity of water = $1.24 \times 10^{-3}$ kg m$^{-1}$ s$^{-1}$

*Solution*

Assuming that the particles are spherical, a force balance gives

$$(\pi d^3/6)(\rho_s - \rho)r\omega^2 = 3\pi\mu d \frac{dr}{dt}$$

so that
$$dt = \frac{18\mu}{d^2(\rho_s - \rho)\omega^2} \frac{dr}{r}$$

The integration of this equation from zero time to time $t$, and between $r = r_0$ and $r = r$, gives (ref. 3 and 7)
$$t = \frac{18\mu}{d^2(\rho_s - \rho)\omega^2} \ln(r/r_0)$$

Thus the equation for determining the diameter of the particles is
$$d = \sqrt{18\mu \ln(r/r_0)/(t(\rho_s - \rho)\omega^2)}$$

For the present problem,
$$\mu = 1.24 \times 10^{-3} \text{ kg m}^{-1}\text{s}^{-1}$$
$$r = 16.50 + 1.00 \text{ cm} = 0.1750 \text{ m},$$
$$r_0 = 0.165 \text{ m},$$
$$t = 360 \text{ s},$$
$$\rho_s - \rho = 1425 - 998 = 427 \text{ kg m}^{-3}$$

and
$$\omega = 2\pi N/60$$
$$= 2\pi \times 360/60 = 37.70 \text{ rad s}^{-1}$$

Thus,
$$d^2 = \frac{18(1.24 \times 10^{-3})\ln(0.1750/0.1650)}{360 \times 427 \times 37.70^2} \text{ m}^2$$

whence
$$d = 2.5 \times 10^{-6} \text{ m}$$

The particles of organic pigment in the pipetted sample therefore have a diameter not exceeding 2.5 μm.

## 13.1.4 Flow rates in continuous centrifugal sedimentation

*Problem*

A centrifuge with a basket which is 500 mm long and has an inside radius of 50.5 mm is to be used to separate crystals from a dilute, aqueous mother liquor. The optimum speed of rotation for the centrifuge is 60 000 revolutions per minute, and the discharge weir is adjusted so that the depth of liquid at the basket wall is 38.5 mm.

The crystals are approximately spherical, and none are smaller than $2 \times 10^{-6}$ m in diameter.

What is the maximum volumetric flow rate of mother liquor that can be processed by this centrifuge if all the crystals have to be removed?

Density of mother liquor = 998 kg m$^{-3}$

Density of crystals = 1455 kg m$^{-3}$
Viscosity of mother liquor = $1.013 \times 10^{-3}$ kg m$^{-1}$ s$^{-1}$
Acceleration due to gravity = 9.81 m s$^{-2}$

*Solution*

The equation (ref. 2, 3 and 7)

$$Q = \frac{d^2(\rho_s - \rho)}{18\mu} \cdot \frac{\omega^2 V}{\ln(R/r_i)} \tag{13.7}$$

which has already been used in Example 13.1.1 can also be used to solve the present problem, providing that the terminal settling velocity of the particles is in the Stokes law range.

The terminal settling velocity ($U_t$) of the crystals is given by

$$U_t = d^2(\rho_s - \rho)g/18\mu \tag{13.8}$$
$$= (2 \times 10^{-6})^2 (1455 - 998)(9.81)/(18 \times 1.013 \times 10^{-3}) \text{ m s}^{-1}$$
$$= 9.835 \times 10^{-7} \text{ m s}^{-1}$$

To check that the use of equation (13.8) is justified, the Reynolds number must be determined.
Thus $Re = dU_t \rho/\mu$
$$= (9.835 \times 10^{-7} \times 2 \times 10^{-6} \times 998)/(1.013 \times 10^{-3})$$
$$= 1.94 \times 10^{-6}$$

Equation (13.7) can therefore be used as $Re < 2$.

Now $\qquad \omega = 2\pi N/60$
$\qquad\qquad\qquad = 2\pi \times 60\,000/60 = 6283$ rad s$^{-1}$

Also, $\qquad r_i = 50.5 - 38.5$ mm
$\qquad\qquad\quad = 12.0 \times 10^{-3}$ m

and $\qquad V = \pi L(R^2 - r_i^2)$
$\qquad\qquad\quad = 0.500\pi (0.0505^2 - 0.0120^2)$ m$^3$
$\qquad\qquad\quad = 3.780 \times 10^{-3}$ m$^3$

Substitution of all known values into equation 13.7 gives

$$Q = \frac{(2 \times 10^{-6})^2 (1455 - 998)}{18 \times 1.013 \times 10^{-3}} \cdot \frac{6283^2 \times 3.780 \times 10^{-3}}{\ln(50.5 \times 10^{-3}/12.0 \times 10^{-3})} \text{ m}^3 \text{ s}^{-1}$$
$$= 0.010 \text{ m}^3 \text{ s}^{-1}$$

The use of equation (13.7) implies that the volumetric flow rate, $Q$, is sufficiently low to provide sufficient time for all particles of diameter greater than $2.0 \times 10^{-6}$ m to travel from the inner surface of the liquid to the basket wall, i.e. the time taken for the collection of the particle in the most unfavourable position when introduced into the centrifuge basket. Any higher volumetric flow rate will not allow sufficient residence time for this to occur.

Therefore the maximum volumetric flow rate of the mother liquor is $0.010$ m$^3$ s$^{-1}$.

## 13.1.5 The separation of two immiscible liquids by centrifugation

### Problem

Cream is to be separated from milk in a centrifuge. The weir for the overflow of cream has been set at 65.5 mm from the axis of the centrifuge, and the outlet radius for the skimmed milk is 78.2 mm from the axis.

Calculate the distance of the cream/milk interface in the centrifuge from the axis.

Density of cream = 867 kg m$^{-3}$
Density of skimmed milk = 1034 kg m$^{-3}$

### Solution

In a centrifuge which is separating two liquid phases, the location of the interface can be determined using a force balance, which leads to the equation (refs. 1, 4 and 5)

$$P_2 - P_1 = 0.5\, \rho \omega^2 (r_2^2 - r_1^2)$$

where $P_1$ and $P_2$ are the pressures at two positions in the centrifuging liquid which are at distances $r_1$ and $r_2$ respectively from the axis of rotation.

If a light liquid phase in the centrifuge bowl has a thickness $r_2 - r_1$, and the radius to the surface of the heavy liquid downstream is $r_3$, then by equating the pressure exerted by the lighter liquid phase at the liquid–liquid interface to the pressure exerted by the heavier phase at the interface we have that (ref. 4, 5 and 8),

$$0.5\, \rho_L \omega^2 (r_2^2 - r_1^2) = 0.5\, \rho_H \omega^2 (r_2^2 - r_3^2)$$

i.e.

$$r_2^2 = (\rho_H r_3^2 - \rho_L r_1^2)/(\rho_H - \rho_L)$$

Substituting into this equation the values given in the problem statement, we obtain

$$r_2^2 = \frac{(1034 \times 0.0782^2) - (867 \times 0.0655^2)}{1034 - 867}\ \text{m}^2$$

whence the distance from the axis of rotation to the cream–milk interface,

$$r_2 = 0.125\ \text{m}$$

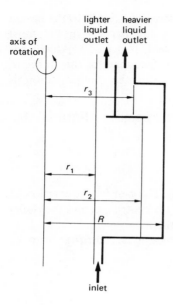

**Figure 13.1**

It should be noted that to solve this problem only the densities of the two liquids have been used to locate the position of the liquid–liquid interface, and no consideration has been given to the relative volumetric flow rates (and hence residence times) of the two liquids in the centrifuge, (see Problem 5).

### 13.1.6 Cyclone separators

*Problem*

A cyclone separator has a diameter of 0.333 m and a height of 1.28 m. The diameter of the circular inlet and outlet both equal 9.50 cm. The cyclone is expected to remove spherical particles of mineral ore, of density 2655 kg m$^{-3}$, down to a particle diameter of $2.0 \times 10^{-6}$ m, from an air stream.
Determine the minimum air flow rate (in m$^3$ s$^{-1}$) into the cyclone.

Density of air = 1.210 kg m$^{-3}$
Viscosity of air = $1.780 \times 10^{-5}$ kg m$^{-1}$ s$^{-1}$
Acceleration due to gravity = 9.81 m s$^{-2}$

## Solution

Theoretically the size of the smallest particle retained by a cyclone, or the minimum air flow rate to retain a given size, can be calculated as follows, but experience shows that these calculations are only approximate as agglomeration of particles, and entrainment, tend to introduce deviations from calculated values (ref. 9).

By considering the two principal forces acting in a cyclone, namely the centrifugal force acting outwards and the frictional drag of the gas acting inwards, the following equation relating the terminal velocity, $U_t$, of a particle to fluid properties and the geometry of a cyclone separator has been derived (ref. 10).

$$U_t = 0.2 A^2 d_o \rho g / \pi z d_t \dot{m} \tag{13.9}$$

Thus, from a knowledge of the terminal settling velocity of a particle the required mass flow rate ($\dot{m}$) of air, and therefore the volumetric flow rate, can be calculated.

If one assumes that the settling is in the Stokes law range, $U_t$ can be found from the following equation. Thus

$$\begin{aligned} U_t &= d^2 g (\rho_s - \rho)/18\mu \tag{13.10} \\ &= (2.0 \times 10^{-6})^2 (9.81)(2655 - 1.21)/(18 \times 1.780 \times 10^{-5}) \, \text{m s}^{-1} \\ &= 3.250 \times 10^{-4} \, \text{m s}^{-1} \end{aligned}$$

To check that the settling is in fact in the Stokes law range, the Reynolds number ($Re$) for the settling particle should be determined.

Thus
$$\begin{aligned} Re &= d U_t \rho/\mu \\ &= (2 \times 10^{-6} \times 3.25 \times 10^{-4} \times 1.21)/(1.78 \times 10^{-5}) \\ &= 4.42 \times 10^{-5} \end{aligned}$$

Since this is less than 2, the use of equation (13.10) to determine the terminal settling velocity of the ore particle is justified.

Now equation (13.9) can be rearranged as follows to give the mass flow rate of the air directly.

$$\dot{m} = 0.2 A^2 d_o \rho g / U_t \pi z d_t$$

The substitution of known values gives

$$\dot{m} = \frac{0.2[\pi(0.095/2)^2]^2 0.095 \times 9.81 \times 1.21}{3.250 \times 10^{-4} \times \pi \times 1.28 \times 0.333} \, \text{kg s}^{-1}$$

$$= 0.0260 \, \text{kg s}^{-1}$$

Thus, volumetric flow rate,

$$\begin{aligned} Q &= \dot{m}/\rho \\ &= 0.0260/1.21 \, \text{m}^3 \, \text{s}^{-1} \\ &= \underline{0.021 \, \text{m}^3 \, \text{s}^{-1}} \end{aligned}$$

### 13.1.7 Efficiency of cyclone separators

*Problem*

The efficiency ($E$) of a certain cyclone separator is related to the diameter ($d$) of the particles in a gas passing through it by the following equation

$$E = 1 - \exp(-4.46 \times 10^4 d)$$

when $d$ is measured in metres.

This cyclone is to be used to remove particles for which the size range is described by a particular form of the Rosin–Rammler distribution function (ref. 11), i.e.

$$x = 1 - \exp(-1.98 \times 10^4 d)$$

where $x$ is the mass fraction of particles having a diameter less than $d$ metres.

Determine the overall efficiency for the cyclone.

*Solution*

If $x$ is the mass fraction of particles of diameter less than $d$, then the mass of particles having diameters between $d$ and $(d + \Delta d) = \Delta x\, W_t$, where $W_t$ is the total mass of the particles in a sample of the gas/solid mixture. Denoting the mass of these particles that is collected in the cyclone by $\Delta W_c$, then

$$\Delta W_c = \Delta x\, W_t E$$

where $E$ is the efficiency of the collection operation with respect to particles in the size range between $d$ and $(d + \Delta d)$.

It follows that the total mass of all the particles collected by the cyclone separator is given by

$$W_c = W_t \int_0^1 E\, dx$$

Now $W_c/W_t$ is the overall efficiency, $E_0$, so that

$$E_o = \int_0^1 E\, dx \tag{13.11}$$

Now both $E$ and $x$ are functions of the particle diameter, $d$. Often these functions are available only in numerical form, so graphical integration of equation (13.11) is appropriate (see Problem 7 for example). However, for the present example

$$x = 1 - \exp(-1.98 \times 10^4 d)$$

and

$$dx = 1.98 \times 10^4 \exp(-1.98 \times 10^4 d)\, dd \tag{13.12}$$

Furthermore,

$$E = 1 - \exp(-4.46 \times 10^4 d) \tag{13.13}$$

*Centrifugal separation operations* 269

Substitution of equations (13.12) and (13.13) into equation (13.11) gives

$$E_0 = \int_{d=0}^{d=\infty} [1 - \exp(-4.46 \times 10^4 d)][1.98 \times 10^4 \exp(-1.98 \times 10^4 d)] \, dd \quad (13.14)$$

Note that the change of variables necessitated a change in the limits of integration, because for $x = 0$, $d = 0$ and for $x = 1$, $d = \infty$.

Equation (13.14) may be conveniently integrated by putting

$$4.46 \times 10^4 = b$$

and

$$1.98 \times 10^4 = c$$

i.e.

$$E_0 = \int_{d=0}^{d=\infty} [1 - \exp(-bd)][c\exp(-cd)] \, dd$$

$$= c \int_{d=0}^{d=\infty} [\exp(-cd) - \exp(-bd)\exp(-cd)] \, dd$$

$$= c \int_{d=0}^{d=\infty} \{\exp(-cd) - \exp[-d(b+c)]\} \, dd$$

$$= c \left[ -\frac{1}{c}\exp(-cd) + \frac{1}{b+c}\exp[-d(b+c)] \right]_{d=0}^{d=\infty}$$

$$= 1 - \frac{c}{b+c}$$

Thus

$$E_0 = \frac{b}{b+c}$$

Substituting back the values of $b$ and $c$, we obtain

$$E_0 = (4.46 \times 10^4)/(6.44 \times 10^4)$$

$$= \underline{0.69}$$

The overall efficiency of the cyclone separator is therefore 69%.

## 13.2 Student Exercises

1  A processing plant uses a gravity settler with a cross-sectional area of 258 m² to continuously remove particles from a process stream. The settler is to be replaced by a centrifuge which has a tubular bowl of inside diameter 0.0445 m and length of 0.184 m. If the centrifuge is adjusted so that the thickness of the liquid layer is 12.2 mm, determine the speed of rotation which it should have.

Acceleration due to gravity = 9.81 m s⁻²

270  *Problems in fluid flow*

**2** A centrifuge with a bronze basket of diameter 31.5 cm is to be run at a speed of 3750 revolutions per minute. If the fluid which is being clarified has a density of 1013 kg m$^{-3}$, and the thickness of the walls of the basket is 10.56 mm, what is the maximum thickness of the liquid layer that should be allowed at the walls?

Density of bronze = 8895 kg m$^{-3}$
Safe working stress for bronze = $55 \times 10^6$ N m$^{-2}$

**3** A solution containing a small number of talc particles in suspension is to be partly clarified by centrifugation. The centrifuge has a tubular bowl which is 18.42 cm long, has an inside diameter of 4.45 cm, and operates at 1600 revolutions per minute. It is adjusted so that the liquid depth on the wall of the bowl is 1.88 cm.

If the centrifuge processes 0.425 m$^3$ of solution per hour, what is the diameter of the smallest particles that will be removed?

Density of solution = 1002 kg m$^{-3}$
Density of talc = 2418 kg m$^{-3}$
Viscosity of solution = $1.018 \times 10^{-3}$ kg m$^{-1}$ s$^{-1}$

**4** When an aliquot of a fungicide dispersion was centrifuged at 5000 revolutions per minute for 55 minutes, 0.175 cm of clear, supernatant liquid appeared. The centrifuge tube contained dispersion to a depth of 5.75 cm, and the surface of the dispersion was 9.25 cm from the axis of rotation.

How long will it take 5% of the dispersion to separate as supernatant liquid under normal storage conditions?

Acceleration due to gravity = 9.81 m s$^{-2}$

**5** An oil has to be separated from water using a centrifuge which rotates at 6325 revolutions per minute. The bowl of the centrifuge is 76.0 cm long and has an inside radius of 7.72 cm. The radius of the weir over which the oil leaves the centrifuge is 1.62 cm. The mixture, which is to be fed to the centrifuge at a rate of 0.782 m$^3$ per hour, contains 20% by volume of oil. It is a requirement that both the oil and the water have the same mean residence time in the centrifuge.
(i) Determine the distance of the oil–water interface from the axis of rotation of the centrifuge.
(ii) Determine the mean residence time of the oil in the centrifuge.
(iii) Determine the radius of the weir over which the water leaves the centrifuge.
(iv) Determine the maximum diameter of droplets of oil held in the water.

Density of oil = 918 kg m$^{-3}$
Density of water = 998 kg m$^{-3}$
Viscosity of water = $1.013 \times 10^{-3}$ kg m$^{-1}$ s$^{-1}$

6   A cyclone separator has a diameter of 0.382 m and is 1.27 m long. It has circular inlet and exit ducts, both of which have the same diameter. The cyclone is expected to process 23.50 m³ of dust laden air per hour, and the particle size cut which is required is $5.0 \times 10^{-6}$ m.

Determine a suitable diameter for the inlet and exit ducts.

Density of air = 1.210 kg m$^{-3}$
Density of dust particles = 1755 kg m$^{-3}$
Viscosity of air = $1.78 \times 10^{-5}$ kg m$^{-1}$ s$^{-1}$
Acceleration due to gravity = 9.81 m s$^{-2}$

7   A flue gas contains cement particles which have the following particle size distribution.

| Diameter, $d$ (μm) | Mass % of particles with diameter less than $d$ |
|---|---|
| 5.1 | 4.0 |
| 10.0 | 9.0 |
| 17.8 | 20.0 |
| 23.0 | 30.0 |
| 34.3 | 50.0 |
| 40.5 | 60.0 |
| 49.5 | 70.0 |
| 70.0 | 80.0 |

Before being released into the atmosphere the flue gas is passed through a cyclone separator which is known to have the following efficiencies for different particle diameters.

| Efficiency | Particle diameter (μm) |
|---|---|
| 0.10 | 2.0 |
| 0.25 | 4.5 |
| 0.35 | 6.0 |
| 0.45 | 7.5 |
| 0.55 | 8.5 |
| 0.65 | 9.7 |
| 0.75 | 12.0 |
| 0.85 | 16.9 |
| 0.92 | 24.0 |
| 0.96 | 40.0 |
| 0.98 | 50.0 |
| 0.99 | 70.0 |

272  Problems in fluid flow

Determine the mass fraction of particles which will be removed in the cyclone.

## 13.3 References

References are abbreviated as follows:

C and R (2) = J. M. Coulson and J. F. Richardson, Chemical Engineering, Vol. 2, 3rd edition, Pergamon, 1980

F et al. = A. S. Foust, L. A. Wenzel, C. W. Clump, L. Maus and L. B. Andersen, Principles of Unit Operations, 2nd edition, John Wiley and Sons, 1980

C. J. G. = C. J. Geankoplis, Transport Processes and Unit Operations, Allyn and Bacon, Inc., 1978

1. C and R (2), page 205
2. F et al., page 625
3. C. J. G., page 577
4. F et al., page 628
5. C. J. G., page 579
6. J. H. Perry (ed), Chemical Engineers' Handbook, 5th ed., McGraw-Hill, 1973, pages 19–88
7. C and R (2), page 207
8. C and R (2), page 208
9. F et al., page 621
10. C and R (2), page 307
11. T. Allen, Particle Size Measurement, 2nd edition, Chapman and Hall, 1975, page 98

## 13.4 Notation

| Symbol | Description | Unit |
|---|---|---|
| $A$ | cross-sectional area of gas inlet to cyclone | $m^2$ |
| $a$ | acceleration of a particle | $m\,s^{-2}$ |
| $b$ | a constant | — |
| $c$ | a constant | — |
| $d$ | particle diameter | m |
| d | differential operator | — |
| $d_o$ | diameter of outlet of cyclone | m |
| $d_t$ | diameter of cyclone | m |
| $E$ | efficiency of a cyclone for particles of a given diameter | — |
| $E_o$ | overall efficiency of a cyclone | — |
| $F_c$ | centrifugal force | N |
| $F_g$ | gravitational force (weight) | N |
| $g$ | acceleration due to gravity | $m\,s^{-2}$ |
| $h$ | depth of fluid on the wall of a centrifuge bowl | m |
| $L$ | length (or depth) of a centrifuge bowl | m |
| $m$ | mass of a particle | kg |
| $\dot{m}$ | mass flow rate of fluid | $kg\,s^{-1}$ |
| $N$ | speed of rotation of a centrifuge | $rev\,min^{-1}$ |
| $P$ | pressure | $N\,m^{-2}$ |
| $Q$ | volumetric feed rate to a centrifuge | $m^3\,s^{-1}$ |
| $R$ | radius of centrifuge basket or bowl | m |
| Re | Reynolds number | — |

| Symbol | Description | Unit |
|---|---|---|
| $r$ | radial distance from axis of rotation | m |
| $r_1$ | radial distance to the inner liquid surface from axis of rotation of a centrifuge | m |
| $r_0$ | radial distance to surface of liquid in a batch centrifuge at $t = 0$ | m |
| $S$ | stress on walls of a centrifuge basket | $N\,m^{-2}$ |
| $t$ | time | s |
| $U_t$ | terminal settling velocity of a particle | $m\,s^{-1}$ |
| $V$ | volume of suspension, or liquid, in a centrifuge basket or bowl | $m^3$ |
| $W_c$ | mass of particles collected by a cyclone | kg |
| $W_t$ | total mass of particles in a sample | kg |
| $\omega$ | angular velocity | $rad\,s^{-1}$ |
| $x$ | mass fraction of particles having a diameter less than $d$ | — |
| $z$ | height of a cyclone separator | m |
| $\mu$ | viscosity of fluid | $kg\,m^{-1}\,s^{-1}$ |
| $\rho$ | density of fluid | $kg\,m^{-3}$ |
| $\rho_b$ | density of bronze | $kg\,m^{-3}$ |
| $\rho_H$ | density of heavy liquid in a liquid–liquid centrifuge | $kg\,m^{-3}$ |
| $\rho_L$ | density of light liquid in a liquid–liquid centrifuge | $kg\,m^{-3}$ |
| $\rho_s$ | density of particles | $kg\,m^{-3}$ |
| $\Sigma$ | capacity factor for a centrifuge | $m^2$ |
| $\tau$ | thickness of centrifuge bowl | m |

# Answers to Exercises

## Chapter 1
1. (i) turbulent, (ii) laminar, (iii) $5.74 \text{ kg s}^{-1}$
2. (i) $-10.42$ Pa, (ii) $1.937 \times 10^{-3}$, (iii) $1629$ N, (iv) $2.686 \times 10^{-3} \text{ kg m}^{-1} \text{s}^{-1}$ (v) $-533.5 \text{ s}^{-1}$
3. (i) $1.402 \times 10^{-3} \text{ kg m}^{-1} \text{s}^{-1}$ (ii) $106.3$
4. $0.04\%$
5. (ii) $2.74\%$ too high
6. (i) $3548$ kW, (ii) $321.8$ kPa, (iii) $3299$ kW
7. (i) $29.27 \text{ m s}^{-1}$ (ii) $28.16 \text{ dm}^3 \text{s}^{-1}$ (iii) $12.05$ kW
8. $24.03$ kW
9. $2.20 \text{ dm}^3 \text{s}^{-1}$
10. $5.12$ cm

## Chapter 2
1. (i) $4.48$ cm (ii) $523.4$ kPa
2. (i) $22.72$ m (ii) $14.84 \text{ g s}^{-1}$ (iii) $237.0$ kPa (iv) $34.21 \text{ g s}^{-1}$ (v) $813 \text{ J s}^{-1}$
3. (i) $124.8$ kPa (ii) $0.569 \text{ kg s}^{-1}$ (iii) $-25\,°\text{C}$
4. (i) $1.056$ MPa (ii) $5.63$ cm
5. (ii) $0.101 \text{ kg s}^{-1}$, $32.71 \text{ dm}^3 \text{s}^{-1}$, $0.531 \text{ m s}^{-1}$
6. $1611$ Pa

## Chapter 3
1. (i) $64.78$ cm, (ii) $9.505$ mm, (iii) $10.66 \text{ cm s}^{-1}$, (iv) $4.023 \times 10^{-2} \text{ N m}^{-2}$ (v) $4.898 \times 10^{-2} \text{ N m}^{-2}$
2. (i) $98.32\%$ (ii) $4.586$ cm (iii) $0.204 \text{ m s}^{-1}$ (iv) $0.1650$ mm (v) $6.754 \text{ N m}^{-2}$ (vi) $7.360 \text{ N m}^{-2}$ (vii) $50.97 \text{ N m}^{-2}$
3. (i) $5.28 \text{ dm}^3 \text{s}^{-1}$ (ii) $-15.95 \text{ s}^{-1}$ (iii) $0.314 \text{ N m}^{-2}$ (iv) $0.598 \text{ m s}^{-1}$ (v) $0.1595 \text{ m s}^{-1}$
4. (ii) $3.151 \times 10^{-3}$ (iii) $0.536$ mm (iv) $2.677$ mm (v) $2.790 \text{ m s}^{-1}$ (vi) $155.1 \text{ s}^{-1}$, $-3.055 \text{ N m}^{-2}$ (vii) $36.65 \text{ N m}^{-2}$

## Chapter 4
1. (i) $5.25$ cm (ii) $30.50$ cm
2. $334.5 \text{ dm}^3 \text{s}^{-1}$
3. $3.64$ cm
4. $45.05$ cm
5. (i) $706.6$ Pa (ii) $2.46$ cm
6. $35.6\%$

Answers to exercises 275

## Chapter 5

1. $0.850 \text{ m}^3\text{ s}^{-1}$, $10.66 \text{ N m}^{-2}$
2. $15.0 \text{ cm}$, $57.6 \text{ m}^{0.5}\text{ s}^{-1}$
3. (i) $0.835 \text{ m}$, (ii) $0.0455$ degrees
4. (i) $2.627 \text{ m}$ (ii) $8.07 \text{ m}^3\text{ s}^{-1}$
5. (i) $0.426 \text{ m}^3\text{ s}^{-1}$, $0.391 \text{ m}^3\text{ s}^{-1}$
   (ii) $0.533 \text{ m}$, $0.564 \text{ m}$ (iii) $24.97 \text{ dm}^3\text{ s}^{-1}$
6. (i) $0.8857 \text{ m}$, (ii) $0.163$ (iii) $0.107 \text{ m}$ (iv) $3.81$ (v) $1.32\%$
   (vi) $87.92\%$ (vii) $3.908 \text{ m}^3\text{ s}^{-1}$
7. $0.570 \text{ m}$, $0.320 \text{ m}$, $0.210$ degrees, $1.053$ degrees
8. (i) $0.524 \text{ m}$, (ii) $1.603 \text{ m s}^{-1}$, (iii) $0.280 \text{ m}^3\text{ s}^{-1}$
9. (i) $1.260 \text{ m}$ (ii) $0.380 \text{ m}$
10. (i) reach b, (ii) $41.36 \text{ m}$ upstream from the junction with reach c,
    (iii) $3.24 \text{ W}$

## Chapter 6

1. (i) $6.668 \text{ m}$, (ii) $1.009 \text{ dm}^3\text{ s}^{-1}$, (iii) $4.03 \text{ cm}$
2. (i) $0.4464 \text{ m}^{0.75}\text{ s}^{-1.5}$, (ii) $4.606 \text{ kW}$ (iii) $12.72 \text{ m}$ (iv) $41.6 \text{ dm}^3\text{ s}^{-1}$
3. $15.9 \text{ dm}^3\text{ s}^{-1}$
4. 38 stages
5. 8 pumps
6. (i) $4.73 \text{ m}$ (ii) $0.198 \text{ dm}^3\text{ s}^{-1}$
7. $2.27 \text{ m}$

## Chapter 7

1. $67.5 \text{ s}$
2. $0.787 \text{ mm}$
3. $4.68 \text{ cm}$

## Chapter 8

1. $76.85 \text{ kPa}$, $4.80 \text{ cm}$, $3.361 \times 10^{11} \text{ m}^{-2}$, $48.04 \text{ }\mu\text{m}$, $0.958 \text{ mm}$
2. (i) $172 \text{ min } 30 \text{ s}$ (ii) $31 \text{ min } 25 \text{ s}$
3. (i) $0.1123 \text{ m}^3$ (ii) $11.7 \text{ dm}^3$ (iii) $1.02 \text{ cm}$ (iv) $0.836 \text{ }\mu\text{m}$ (v) $1.12 \text{ mm}$
4. $69 \text{ min } 52 \text{ s}$
5. (i) $17.97 \text{ g}$ benzoic acid per kg slurry (ii) $23.48 \text{ dm}^3\text{ s}^{-1}$
6. (i) $67.71 \text{ kg}$ (ii) $1.06 \text{ m}^3$

## Chapter 9

1. The experimental results can be fitted to the published plots of drag coefficient versus Reynolds number for spherical particles, and cover Reynolds numbers ranging from $5.5 \times 10^{-7}$ to $1 \times 10^4$.

276  *Problems in fluid flow*

2  $17.3 \text{ m s}^{-1}$
3  Figure 9.3
4  $1.49 \text{ kg m}^{-1} \text{s}^{-1}$
5  $6.7 \times 10^{-6}$ m
6  (i) $0.0129 \text{ m s}^{-1}$  (ii) 4.14 ms  (iii) 0.03 mm
7  135.9 m

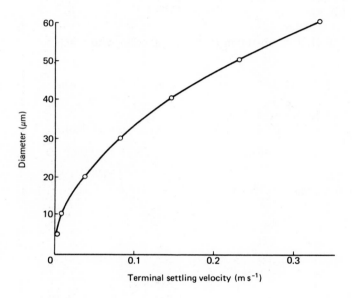

**Figure 9.3**

## Chapter 10
1  $1.69 \times 10^{-5} \text{ m s}^{-1}$
2  $35900 \text{ kg h}^{-1}$
3  7.5 m
4  Figure 10.6
5  $45 \times 10^{-6}$ m
6  84%
7  (i) 50%  (ii) 11%

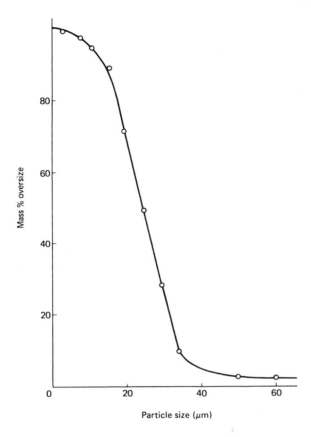

**Figure 10.6**

## Chapter 11
1. $U = 3.36 \times 10^{-3} e^{5.0}$ m s$^{-1}$
2. (i) 1123 kg (ii) 0.73
3. (i) 0.70 mm (ii) 0.34 m s$^{-1}$
4. 0.26 m
5. 15 600 N m$^{-2}$
6. (i) Figure 11.2
   (ii) In the streamline region ($Ga < 3.6$) the ratio of $U_t/U_{mf}$ is about 65, whereas in the turbulent region ($Ga > 10^5$) the ratio is about 10. As fluidisation occurs at superficial velocities between $U_{mf}$ and $U_t$, the range of velocities over which particulate fluidisation can be obtained is very much greater in the streamline region than the turbulent region.
7. $1.48 \times 10^{-3}$ m s$^{-1}$
8. 2790

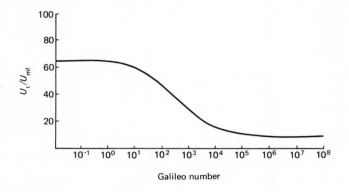

**Figure 11.2**

## Chapter 12

1. (i) No  (ii) Yes
2. 4.5 m s$^{-1}$ (superficial linear velocity)
3. $9.7 \times 10^{-3}$ m$^3$ s$^{-1}$
4. (i) 0.782 N m$^{-2}$ per metre length of pipe  (ii) 11.0 N m$^{-2}$ per metre of pipe  (iii) 82.6 N m$^{-2}$ per metre length of pipe
5. (i) 14.5 N m$^{-2}$  (ii) 58.5 m  (iii) 59.0 m
6. 7.9 m s$^{-1}$

## Chapter 13

1. 23 600 rev min$^{-1}$
2. 83 mm
3. $19.1 \times 10^{-6}$ m
4. $2.40 \times 10^7$ s
5. (i) 3.74 cm  (ii) 62.6 s  (iii) 1.88 cm  (iv) $2.5 \times 10^{-6}$ m
6. 10.2 cm
7. 0.88

# Appendix

Figures A1 and A2 are reprinted with permission from J. M. Coulson and J. F. Richardson, Chemical Engineering, Vol. 1, 2nd edition, 1966, Pergamon Press Ltd.

**Figure A1**
The chart of $\phi$ versus $Re$

**Figure A2**
The chart of $\phi . Re^2$ versus $Re$